History of The Improved Short-Horn or Durham Cattle

And Notes On The Kirklevington Herd by Thomas Bates

by Thos. Bates and Thos. Bell

with an introduction by Jackson Chambers

This work contains material that was originally published in 1871.

This publication is within the Public Domain.

This edition is reprinted for educational purposes and in accordance with all applicable Federal Laws.

Introduction Copyright 2017 by Jackson Chambers

Self Reliance Books

Get more historic titles on animal and stock breeding, gardening and old fashioned skills by visiting us at:

http://selfreliancebooks.blogspot.com/

Introduction

I am pleased to present another title in the "Cattle" series.

The work is in the Public Domain and is re-printed here in accordance with Federal Laws.

As with all reprinted books of this age that are intended to perfectly reproduce the original edition, considerable pains and effort had to be undertaken to correct fading and sometimes outright damage to existing proofs of this title. At times, this task is quite monumental, requiring an almost total "rebuilding" of some pages from digital proofs of multiple copies. Despite this, imperfections still sometimes exist in the final proof and may detract from the visual appearance of the text.

I hope you enjoy reading this book as much as I enjoyed making it available to readers again.

Jackson Chambers

THOMAS BATES.
(From a Portrait by Sir William Ross.)

TABLE OF CONTENTS.

INTRODUCTION.

The Defects of Agricultural Literature—Animal Food Appreciated—Mr. Bates on Commerce and Agriculture—Mr. Bakewell, the First Improver of Stock—Mr. George Culley takes up Mr. Bakewell's Labours and Honours—Mr. John Bailey's Nativity, Career, and Views—Mr. George Culley's Principles of Breeding Stock followed by Mr. Bates—Long-horns and Short-horns—The Author's Position and Object—Messrs. Colling, their Judgment and Herd—The First Herd-Book and Mr. Bates—Mr. George Coates, Mr. Mason, Mr. Hutchinson, Mr. Berry, and their Connection with Improved Short-horn Breeding—The Author's Special Knowledge, and the Value of Mr. Bates's Opinions and Predictions .... 1 to 13.

CHAPTER I.

PART I.

EARLY OBSERVATIONS, REFLECTIONS, AND MATURED VIEWS.

Mr. Bates, Mr. Wastell, of Burdon, and the Importance of Agricultural Improvement—Messrs. H. and W. Taylor, of Chirston Bank, Opulent Northumberland Graziers—Extra Value of Improved Cattle in the Yorkshire and Durham Markets—Mr. Harrison at Barmpton—C. Colling's Bull, Favourite—Mr. Wastell Bred Mr. Jolly's Bull (337)—Mr. Maynard's Cow, Favourite, and the Duchesses, Daisys, and Cherries—Mr. Wastell and an Authentic History of Mr. Jolly's Bull, Jolly—Mr. Jobson, Turvelaws, Wooler, and another Romantic Incident in the life of Jolly—Mr. Wastell and his Influence on the Young Mind of Mr. Bates—Mr. Bakewell's Improved Leicesters, and a Black-faced Tup—How Animals Breed Back—Short-horns Great Milkers, Grazers, and Meat Producers—Various Yield of Milk per Diem—Mr. Wastell's Cow, Barforth, gave 36 Quarts—Mr. Bates on the

Origin of Short-horns—Mr. Michael Dobinson, the Isle, Sedgefield, the Founder of the Best Tribes—The Durham Agricultural Society, and its Influence on Breeding Short-horns—Mr. John Erasmus Blackett and the Studley Short-horns—Sir James Pennyman, his Tenantry, how he encouraged them, and the celebrated Studley Bull—Mr. Lakeland's, Mr. William Barker's, and Mr. John Brown's Bulls were by the Studley Bull—Sir Hugh Smithson, and the Tradition of the Stanwick Short-horns—Short-horn History to 1783..14 to 28.

PART II.

THE PROGRESSIVE DEVELOPMENT OF THE REAL
SHORT-HORNS.

The Durham Society's Prizes to 1797—Mr. C. Colling visited Mr. Bakewell—Mr. Bates on Mr. C. Colling's Judgment—The Cow Duchess—The Grey Bull Hubback—Mr. Bates, Mr. Hustler, Mr. C. Colling, and a Discussion on "Handling"—Styford, the Grey Bull Hubback, and Belvedere (1706)—Duchess, a Stanwick Cow—Mr. C. Colling and Mr. Booth—The Bull Hubback sold for One Guinea and 30 Guineas—Hubback, Mr. Bates's Description of—Hubback and the Herd Book—The Herd Book an Abortion: Erroneous Entries—The late Best Cow in England—Messrs. Colling acknowledge their Cardinal Error—Messrs. Culley's Herd—Hubback Cows and a Favourite Bull, the source whence fine Short-horns sprung—Hair and Handling, or quality of flesh the only trustworthy indication of purity...........................29 to 41.

PART III.

THE DEVELOPMENT OF REAL SHORT-HORNS
CONFIRMED.

Short-horn Breeders Discouraged—Imperfect Judges—Bolingbroke and a Red-polled Scotch Cow—The "Fool's Piece"—The Wonderful Ox—A more wonderful Heifer—The Messrs. Colling's Great Service—Prophetic Vision, and Valuing a "Chance" at 2,000 Guineas—The Earl, the Issue of the "Chance" and his Premature Death—The Earl's Blood Preserved—Hubback's Female Issue, and Belvedere the Foundation of the Bates Short-horns—Mr. Bates highly appreciates Duchess No. 1 in the Herd Book—The Yarborough Bull's Stock Good—A Confession by Mr. C. Colling, and Second Hubback's Handling—Major Rudd a Wrangler and no Judge—Hubback and Second Hubback the only True Ancestors of English Short-horns—Mr. Bates Despairs of Short-horn Breeders—Hereford and Devon Men better Judges—Mr. Mason's Confession—The Great Milking Qualities of the Bates Cows Demonstrated ....42 to 52.

PART IV.

THE PEDIGREES OF THE REAL SHORT-HORNS GIVEN IN FULL.

Hubback bought by Mr. C. Colling in 1785—To what Mr. Bates's success is entirely due—Mr. C. Colling committed another Great Error—Dicky Barker's Bull Described—Lady Maynard not a True Breeder—Bolingbroke's Issue not Uniform—Favourite's Issue a trifle Coarse—Authentic Documentary History of the Pure Short-horns: Mr. Alexander Hall's Hand and Seal—Dicky Barker's Bull—Mr. C. Colling bought his First Duchess Cow for £13 in 1784—Lord Bolingbroke sold for 70 guineas in 1797—Johanna, the Lame Bull, Styford, Cupid, Yarborough, Venus, North Star, Comet, Stick-a-Bitch, Dolly, Princess (by Hubback), Princess Heifer, Foljamb, Daisy, and several other Notable Animals (Male and Female) Named and Described, and their Owners and the Date of their Sales, and other Particulars Recorded ......52 to 63.

CHAPTER II.

PART I.

WHAT MR. BATES CONSIDERED A REAL SHORT-HORN, WITH AMPLE COLLATERAL EVIDENCE IN SUPPORT OF HIS JUDGMENT.

Pure Short-horns—Mr. George Culley on Purity and Quality as instinctively appreciated by Handling—Mr. Culley and Mr. Bakewell—Mr. Bates on Decisions at Agricultural Shows; his Success at York in 1838—Short-horns should be Shown in Families—Belvedere, Waterloo, Angelina 2nd, and Young Wynyard—Mr. C. Colling and Sir H. V. Tempest, Bart., and Princess and Wellington

—More Pedigrees in detail, and Young Wynyard in Ireland—Authentic Documentary Proofs—Twin Heifers Breeding—Judges at Shows overlook Milking Qualities and heed only Fatness—Locomotive went to Kentucky—The Oxford Premium Cow bred Duke of Wellington, which went to Troy, New York—Heavy Weights of Pure Short-horns accounted for—Pure Short horns will Improve from Generation to Generation—Mr. Bates begun Weighing Food and Measuring Cattle, which practice he recommends to all Breeders—Tyneside Short-horns Improved between 1785 to 1795..64 to 76.

PART II.

THE QUALITIES OF MONGREL AND PURE-BRED SHORT-HORNS DISCUSSED, AND THE VALUE OF EACH COMPARED.

Coarse-beefed Short-horns Described—Mr. Culley's Evidence as to the Introduction of Coarse Dutch Animals—Hubback's Influence on the Issue of Coarse Cows—Durham was not free from the Importation of these Inferior Stock—Mistake by Dairymen in regard to the Richness or Quality of Milk—Consumption of Food by Different Animals tested by Mr. Bates—St. John's Delicate Constitution further Proved—The Loss of Ketton Appreciated—The Earl made a Great Mark—The Three Tribes to which the Bates Short-horns Owe their Excellence—Mr. C. Colling and the Introduction of the Galloway Strain to his Herd of Short-horns; the Effect being "Hard Flesh, Bad Wiry Hair, and No Milkers"—The "Fool's Piece" again—One Cross may be Superseded, but Two Crosses never—The Value of this Knowledge—Judges at Shows Incompetent, or Worse..................................76 to 83.

PART III.

MR. BATES COMMENDS HIS PURE HERD TO POSTERITY.

Gambier and Belvedere purchased—The Value of Pure Short-horns of 1845—The principle which guided Mr. Bates; he studied the Fame of his Herd more than his Fortune in Life—Mr. Bates on Crossing Breeds—The Highland Herds and Pure Short-horns—Highlanders made more Hardy by the Introduction of Short-horn Blood—Stock that will not bear the test of Ordinary Hardships must be Unprofitable—Mr. Bates Weeded his Herd of Faulty Cows, as for

Abortion and other Defects—Bolstering up Degenerating Families—What "a Painter without Principle" may do—The Value of Pedigree, and in what this Value consists—Mr. Bates's Object in Writing this History—Hubback, Strawberry, and Strawberry's sister—The Herd Book begun and continued with False Entries—Mr. Coates's Confession, and his Poverty his Excuse—Lily and Fortune Objectionable Cows, and the St. John Bull and his Weak Descendants—Jupiter Coarse and Ill-bred—Mr. Mason admitted losing £20,000 by his Herd, which is Fully and Intelligibly Accounted for by Mr. Bates—Mr. Bates Insulted for being Candid and Truthful; his Assailant Apologizes; why this is referred to—The Galloway Scotch Blood again noticed in Connexion with Bolingbroke—Mr. Bates bought the First Cow Mr. C. Colling sold for 100 Guineas—Comet Valued at 1500 Guineas when 7 Years of Age; Died 1815—Ketton Improved the Comet Issue....84 to 101.

MEMOIR OF MR. BATES,

WITH HIS PERSONAL EXPERIENCE AND RECOLLECTIONS DURING HIS PURSUIT OF AGRICULTURE, AND MORE ESPECIALLY IN REGARD TO HIS PURE HERD OF SHORT-HORNS.

PART I.

Scientific Breeding; a Newspaper Critique; what does it all mean? or what is it all about?—Mr. Bates's Early Life; where and when Born; from whom Descended; had Feeble Health in Early Years; where Educated—Mr. Bates, when young, elected to enter the Church, but his father, though a good Churchman himself, had (for reasons given) a great repugnance to his son's wish—Mr. Bates's Schoolfellows and Fellow-Students, and with whom he formed a life-long Friendship—The Blayney Family—The Family Estate left to a Stranger in blood ....102 to 111.

PART II.

Mr. Bates's Agricultural Career began at Aydon Castle—The Families of Surtees, Bates, and Cook, and the intimacy of the Culleys with Mr. John Bates and his Sons—Mr. George Culley and Arthur Young—Teeswater Landowners and Occupiers; their character, and Mr. George Bates's intimacy with them—The Messrs. Culley, Mr. George Bates, and Major Surtees—Messrs. Culley's Rent £10,000 a year—Mr. George Culley's judgment criticised, but not weakened—Mr. Bates a visitor to, but not a pupil of, Mr. George Culley, who, however, ultimately became the model agriculturist whom Mr. Bates followed—Mr. George Culley died in 1817—Mr. Bates began Farming before of age; his earnestness, energy, and successful treatment of different soils—Secret Tenders in Letting Farms; when inaugurated—Tyneside Farmers in 1800 and their Short-horn Cattle—Mr. Jopling, of Styford, Mr. Wastell, Messrs. Colling, and Mr. Donkin, of Sandhoe—What Landowners had the best breeds of Cattle—Mr. Bates enters upon Halton Castle, and his surrounding Influences—Mr. Bates and the "Wonderful Ox," and Mr. Maynard's Steers and Heifers—Kyloes in Northumberland; their handling and fatting nature as a pure breed, and when crossed with Short-horns—Weights of remarkable Oxen—Mr. Harbottle on Mr. Bates's Beef as sent to the Smithfield Club Dinner—Mr. George Culley and Mr. Bailey on Agricultural Societies—Mr. John Grey's (Dilston) Assumptions......112 to 125.

PART III.

Mr. Bates's address to the Board of Agriculture.......... 125 to 135.

PART IV.

Mr. Bates as an Improving Agriculturist and his intimacy with the Leading Men of his Time—Mr. Curwen's Correspondence—Mr. Mackenzie's Correspondence and Death—Mr. Trevelyan's Letters on an Experimental Farm and Agricultural Chemistry...136 to 150.

PART V.

Mr. Bates wanted the Security which Belongs to Owners of Land—Purchased Kirklevington—Lord Althorp and Mr. Bates and the Duchess Cattle—Mr. John Grey and Star, North Star, and Sparkles—Lord Althorp's Correspondence on Short-horns, and the Politics of his day—Mr. Champion bought Stock of Messrs Colling, and Lord Althorp on his Herd—Lord Althorp's and Mr. Grey's Herds—Lord Althorp's view to the "Main Chance"—Lord Althorp and Mr. Bates become Less Friendly, and the Reason Why—Lord Althorp and Mr. Bates were Actuated by Different Motives; the Result being the Former was not a Successful Breeder...150 to 164.

PART VI.

Mr. Bates on the effect of a Fall in Prices after the American War—Lord John Russell a Protectionist—Mr. Bates on Politics—Agricultural Questions—Sir James Graham's, Lord Althorp's, and Sir Thomas Lethbridge's Correspondence on ditto—Free-trade Prices of Corn resulted in Land being put down to Grass—Mr. Bates an Employer of Labour; on the Poor Law Board; and the Turnpike and Highway Question—On the Improvement of, and the Planting of, Wastes and Morasses—Kyloes on improved Waste Lands—Swedish Gentleman's Surprise at our Naked Hills—Mr. Bates on Agricultural Machinery; he advises, in 1838, upon an American Reaping Machine—Mr. Fawcett's Reminiscences of Halton............ .... 164 to 180.

PART VII.

Mr. Ramsey's Tyneside Herd—Mr. Bates's First Bulls, and other interesting matters—The Ketton Herd and Mr. C. Colling's anxiety—The Barmpton Herd, and Mr. Bates on a Pure Love of Money—Mr. Bates at Home...................... ....180 to 193.

PART VIII.

The Alloy Blood fully discussed—The "Booth Men" and Mr. Thornton's *Circular*—Mr. Carr not a good advocate, but denounces his own theory—The general custom and value of "Handling"—Mr. Bates on In-and-in Breeding................. ....194 to 202.

PART IX.

Mr. Bates as an Instructive Friend—Mr. Rhodes on his own Stock, and the appreciation of Mr. Bates—Mr. Bates on Pedigree—Mr. Coates, Mr. Whittaker, and the Original Herd Book in MS.—Remarks by Mr. Coates on the Duchesses, and the late Mr. H. H. Dixon's partiality—Lord Althorp on the Herd Book—The Almighty Dollar—Mr. Bates and Mr. Grant Duff on Stock-pedigrees.. ....203 to 216.

PART X.

Cattle never Doctored—Libel on 2nd Hubback Refuted—Loss by Epidemic—Intercourse between Breeder and Butchers—A. Maynard: his Stock and Character—Kenyon College—American Skill and Wisdom—The Value of the Duchess Tribe—Sales at America and Tricks in the Trade—American Cow—Appreciation of the Kirklevington Cattle in America—Shows and Prizes...217 to 233.

PART XI.

Durham Agricultuaal Society—Weighing and Measuring Cattle—Tyneside Society—Working Classes: attempts of the Duke of Northumberland to Improve—Crowley's Crew—Manufacturers supporting their own Poor and Schools—Success of Mr. Bates—Envy—Way to manage him—Short-horns in Northumberland—The Messrs. Jobson—Training for Shows—Booth and Crofton—Ruinous consequences—Mr. Bates again exhibits—Yorkshire Society, 1838..................................233 to 246.

PART XII.

Mr. Bates and Judges at Shows—American Opinions and Sympathy—John Grey on Judges—Arthur Young's Farming—Mr. Bates's Rejected Cattle—Oxford Show and Victory—Road and Sea Journey there—Daniel Webster—Political Farming and Banking—Idea on Exhibiting Families of Cattle—Cleveland Lad—Cambridge Meeting—Opinion Vindicated at Liverpool—Hull Meeting and Success—The Victory at York over the Booth Best—Necklace and Bracelet—Their Pedigrees and Victories by Carr—Ceased to Exhibit—Fat Breeding Stock and its Results—Calves like Cats..................................246 to 262.

PART XIII.

Royal Society and Breeding Stock—Rules Neglected—Pandering to the Masses—Wonderful Thanks and Credit to Mr. Booth—Mr. Bates's Exertions—Judges at Shows—Mr. Bates protests—Red Tape—Letter of Appeal to the Members of the Society—Views adopted by Americans—Booth Beef *versus* Milk—Mr. M'Combie, M.P.: his opinion—Best Cattle not now Exhibited—Neglect of Duchesses 34th and 51st—Foreigners Misled by Shows and Training—Its effect—Holland Cattle—Short-horns in Germany—Mr. Carr's results of Alloy-blood—Poor Relations, &c.....262 to 277.

PART XIV.

The Prize Stock of Mr. Bates's Breeders—Spraggon's Herd—J. Wilson's Prize Bull—Lord Althorp's Brindled Black Nose—Chrisp's Herd—Wynyard Herd—St. Alban's—Princess Tribe by Wood's Herd degenerated—Pedigree made clear—Messrs. Stephenson, Greenwell, and Harrison, clear accounts—Belvedere purchased—Dukes 224 and 226—Sheep destroyed, and the Law—Trial of New Tribes—The six Kirklevington Tribes, &c., &c..277 to 296.

PART XV.

Intended to retire in 1851—Mr. Pippet and Appreciation of Stock—Fortune—Opinions on Stock—Ingratitude and Anonymous Slanders—American Appreciation—Humbug—Father and Mother, Deaths, Relations, and Old Reminiscences—Engaged to be Married—Consumption—Advice of Scotch Legal Judge—Obituary notice in *Farmers' Magazine*—Death—Monuments ....297 to 306

PART XVI.

Kirklevington Sale—Particulars—Notice in *Farmers' Magazine*—Why the Prices go Low—Comments on Lord Ducie and Jonas Webb—Racing Phrase—Lord Ducie's anxiety in Breeding—3rd and 5th Duke of York—Mr. Tanqueray—Tortworth Sale and the Americans—Col. Gunter—Marquis of Exeter—Lord Feversham, his judgment and the Bates Blood—Willis's Rooms Sale—Importation from America—*Punch* on the Windsor Sale—History of Divers Animals—Grand Duchesses, &c.......307 to 339.

PART XVII.

Portraits—Mr. Bates—Duchess, 1804—Ketton—Belvedere Pedigrees—Elvira—Duchess 34th—Duke of Northumberland; sums offered for him—Wetherby Herd and Duchesses—Duchess 77th—94th and Portrait—Pedigree—My own experience and taste—Scarborough Exhibition—Mr. Cochrane's purchases for Canada—Lord Dunmore—8th Duke of York—Pedigree—Purchases by Myself and Friends—Loss of Cattle—My own Herd—Intended Sale—Conclusion.. ....339 to 357.

HISTORY OF SHORT-HORNS.

INTRODUCTION.

The Defects of Agricultural Literature—Animal Food Appreciated—Mr. Bates on Commerce and Agriculture—Mr. Bakewell, the First Improver of Stock—Mr. George Culley takes up Mr. Bakewell's Labours and Honours—Mr. John Bailey's Nativity, Career, and Views—Mr. George Culley's Principles of Breeding Stock, followed by Mr. Bates — Long-horns and Short-horns — The Author's Position and Object—Messrs. Colling, their Judgment and Herd—The First Herd-book and Mr. Bates—Mr. George Coates, Mr. Mason, Mr. Hutchinson, Mr. Berry, and their Connection with Improved Short-horn Breeding—The Author's Special Knowledge and the Value of Mr. Bates's Opinions and Predictions.

One of the greatest authorities on agriculture in all its branches during the last generation made use of this pertinent expression:—"It has been matter of surprise to many that none of our countrymen have hitherto attempted to write a treatise expressly upon what farmers call stock."* This was the view, as will be seen below, of Mr. George Culley, a Northumberland farmer, in 1786.

The student in agricultural literature will find his surprise increased in the present day when he views shelves loaded with volumes on agriculture, but with hardly any on live stock. Mr. Culley has collected, I believe, all the writings of the ancients† on cattle, and recorded

---

* "Obervations on Live Stock, containing Hints for Choosing and Improving the Best Breeds of the most useful kinds of Domestic Animals, by George Culley, 1786."

† Varro, Columella, Mago, Palladius, Virgil, &c.

every reference, excepting, perhaps, to the urus or wild bull. This race still exists in Lithuania in its natural state.

The production of beef and mutton did not enter into the speculations of the ancient writers or the calculations of agriculturists; but in the last century animal food became more important than the profits derived from the use of cattle for draught purposes, or for the produce of the dairy.

The late Mr. Bates contended that commerce and agriculture ought to go hand in hand. The markets for animal food opened out by the increase of the population of London, the extension of manufactories in the midland counties, the shipping and coal trades of the northern ports, and other modes of developing the physical resources of the country, stimulated the English agriculturists, and live stock became the most important part of all improved farming. We must not overlook the fact that animal food was not the staff of life of the ancient inhabitants of southern climes, but in the damp and northern districts it is essential for sustaining strong bodily exertion.

In going through "The Annals of Agriculture," by Arthur Young, commenced about 1780, and "The Transactions of the Society for the Encouragement of Arts, Manufactures, and Commerce," from 1783, it is a curious and instructive study to see the various imperfect steps taken for the improvement of cultivation, stock, and implements.

Mr. Bakewell, of Dishley, is allowed by every one to have been the first person who systematically improved the breed of live stock with a view to the production of animal food. In sheep his success was complete during his time, and experience since has given a permanent character to his principles of selection. His improved long-horn cattle also spread much in the midland counties, and Mr. Fowler, of Oxfordshire, at his sale, in 1791, obtained very large prices, several bulls being sold for above 200 guineas; while at Mr. Paget's sale, in 1793, a bull bred by Mr. Fowler brought 400 guineas, showing the estimation in which well-bred cattle were then held. Mr. Bakewell carefully concealed his proceedings, and, except by the observation of results, his experience was lost to the world. In-and-in breeding, however, was no doubt his secret. It is also very curious to read Arthur Young's account of the Tup Club, a society which was established in the midland counties in Mr. Bakewell's lifetime, to secure for the members all the benefit to be derived from the improved breeding of sheep. No life of Mr. Bakewell was ever published, and unfortunately he is only known in history by a short obituary notice in the *Gentleman's Magazine.*

After Mr. Bakewell, the name of George Culley has always been placed as the next highest authority on live stock. What he did for improving the breeds of sheep in Northumberland is a matter of history. (See "Library of Useful Knowledge.") He and his brother Matthew have been mentioned as "the great veterans and leaders in agricultural improvement." They, however, have had no historian. But the impetus given to agricultural progress by them, and the rapid and astonishing improvements which followed in the North of England and in Scotland, created a national wish for some government acknowledgment of the importance of Agriculture to the State, and for the appointment of a Minister of Agriculture, as well as one for Trade and the Colonies, that the same improvement might be made in other parts of the Kingdom. But as this suggestion was not adopted, the result was the establishment of "The Board of Agriculture and Internal Improvement," to which there was not attached any official or government authority. Sir John Sinclair was the president, and Arthur Young the secretary. Probably the only records now existing of the works of the board are the views or reports of the different counties, drawn up at their request, and for their consideration. No reports contain more information or suggestions for improvement than those for the four northern counties.

The agricultural surveys of Northumberland and Cumberland were by J. Bailey and George Culley, and the survey of Durham by J. Bailey.

John Bailey was a native of the banks of the Tees, and had in early life been a mathematical tutor, and became a land surveyor and agent when the commons and waste lands in Durham and Northumberland were divided and enclosed. He was afterwards resident agent at Chillingham, for the estate of the Earl of Tankerville. Both works are full of interest on all mechanical, mineral, and commercial subjects connected with the two counties, and they anticipate all that has ever been done by the legislature for the improvement of landed properties. The chapter on Political Economy in the survey of Durham would not discredit some of the modern professors, and there is the first indication of the possibility of making general railways, and of using the gas which was formed in the distillation of coal for general illuminating purposes. The chapters on live stock, in both volumes, are instructive; that for Durham has, in fact, been used as the History of Short-horn Cattle; all that was known of them being from Mr. G. Culley's information. The survey of Durham was mostly written in 1796, although the publication was delayed until 1810, and J. Bailey, in speaking then of farmers, says:—" In this, as well as in every other district I am acquainted with, the occupiers of large farms have been the first to make improvements—to introduce new implements, new modes of culture, and improved breeds of live stock. It is men of education and superior intelligence who travel to examine the cultivation of distant counties, and improved breeds of cattle, sheep, and other animals, and who have capital to carry into effect whatever they may think will improve their own districts. Messrs. Culley and Charge were the first that led the way, and they have been followed by Messrs. Colling, Mason, Taylor, Nesham, Seymour, and many others, by whose exertions and judicious selection this district will be lastingly benefited."

The principles of breeding stock, and the valuable properties of the different kinds and races, are so clearly laid down by Mr. G. Culley, that I think it best at once to quote them from the work previously named. Mr. Bates always acted on them, and in all his statements and conversation constantly referred to them.

In mentioning the short-horned or Dutch cattle, George Culley writes, in 1786, as follows:—

"This breed, like most others, is better and worse in different districts; not so much I apprehend from the good or bad quality of the land, as from a want of attention in the breeders. In Lincolnshire, which is the furthest south that we meet with any number of this kind of cattle, they are, in general, more subject to lyery,* or black flesh, than those bred further north; and in that rich part of Yorkshire, called Holderness, they are much the same as those south of the Humber, of which we have been speaking. It is probable that they had either stuck more to the lyery black-beefed kind, than their more northern neighbours, at that unfortunate period, when they were imported from the Continent, or that the latter had seen their error sooner. But, from whatever cause this happened, it is a fact, that as soon as we cross the Yorkshire Wolds northward, we find this breed alter for the better; they become finer in the bone, in the carcase, and, in a great measure, free from that disagreeable lyery sort

* The term "lyery" is explained in the following extract from Mr. George Culley's "Observations on Live Stock, &c.:"—"I remember a gentleman of the county of Durham (Mr. Michael Dobinson), who went in the early part of his life into Holland to buy bulls. Those he brought over were of much service in improving the breed; and this Mr. Dobinson and neighbours, even in my day, were noted for having the best breed of short-horned cattle, and sold their bulls and heifers for very great prices. But afterwards some other persons of less knowledge going over, brought home some bulls that in all probability introduced along that coast the distinguishable kind of cattle well known to the breeders adjoining the Tees by the appellation of *lyery* or *double-lyered*, that is, black-fleshed; for, notwithstanding one of these creatures will feed to a vast weight, and though fed ever so long, yet will not have one pound of fat about it, neither within nor without, and the flesh (for it does not deserve to be called beef) is as black and coarse-grained as horse-flesh."

which has brought such an odium upon this (perhaps) *most valuable breed*. When you reach that fine country on both sides the River Tees, you are then in the centre of this breed of cattle; a country that has been long eminent for good stock of all kinds; the country where the Dobinsons first raised a spirit of emulation amongst the breeders, which is still kept up by Mr. Hill, the Messrs. Charge, the Messrs. Colling, Mr. Maynard, &c., &c.

"The great obstacle to the *improvement* of domestic animals seems to have arisen from a common prevailing idea amongst breeders, that no bull should be used in the same herd more than three years, and no tup more than two; because (say they), if used longer the breed will be *too near akin*, and the produce will be *tender, diminutive*, and liable to *disorders:* some have imbibed the prejudice so far as to think it *irreligious;* and if they were by chance in possession of the best breed in the Island, would by no means put a male and female together that had the same sire, or were out of the same dam. But, fortunately for the public, there have been men, in different lines of breeding, whose enlarged minds were not to be bound by vulgar prejudice, or long established modes, and who have proved by many years' experience, that such notions are without any foundation."

"Mr. Bakewell has not had a cross (from any breed than his own) for upwards of twenty years; his best stock have been bred by the nearest affinities; yet they have not decreased in size, neither are they less hardy or more liable to disorders; but, on the contrary, have kept in a progressive state of improvement."

"This mode has also been frequently practised in breeding the best dogs and game-cocks. A certain gentleman, who produced the best pointers in the North of England for many years, never bred from any other than his own, because, he said, he could not find better to cross them with; and I am informed, from good authority, that a breeder of game-cocks, who was very successful, would never allow his breed to be contaminated by crossing with others, and to this precaution he attributed all his superiority."

"But one of the most conclusive arguments that crossing with different stock is not necessary to secure size, hardiness, &c., is the breed of wild cattle in Chillingham Park, in the County of Northumberland. It is well known these cattle have been confined in this park for several hundred years, without any intermixture; and are, perhaps, the *purest breed* of cattle of any in the kingdom. From their situation and uncontrolled state, they must indisputably have bred from the nearest affinities in every possible degree; yet we find these cattle exceedingly hardy, healthy, and well formed, and their size, as well as colour, and many other particulars and peculiarities, the same as they were five hundred years since."

"From these instances it appears there can be *no danger* in breeding by the nearest affinities, provided they are possessed in a *superior degree* of the qualities we wish to acquire; but if not possessed of these, then we ought to procure such of the same kind as have, in the most eminent degree, the valuable properties we think our own deficient in. It is certainly from the *best males and females* that *best breeds* can be obtained or preserved. To breed in this manner is undoubtedly right so long as *better males* can be met with, not only amongst our neighbours, but also amongst the most *improved breeds* in any part of the island, or from any part of the world, provided the expense does not exceed the proposed advantage; and when you can no longer, at home or abroad, find *better males* than your own, then, by all means, breed from them; whether horses, neat-cattle, sheep, &c., for the same rule holds good through every species of domestic animals; but, upon no account, attempt to breed or cross from *worse* than your own; for that would be acting in contradiction to common sense, experience, and that well-established rule—"That best only can beget best," or, which is a particular case of a more general rule, viz., that "Like begets like."*

"On this simple axiom the whole mystery of improving stock seems to depend, and, like many other valuable truths, has been neglected, most probably for its simplicity, and other modes pursued as whim or fancy directed, without either reason or experiment to support or even give the least colour of plausibility to the practice."

After speaking of Mr. Bakewell's long-horns, Mr. Culley says, "There is little doubt but a breed of short-horn cattle might be selected *equal*, if not *superior*, to even that very kindly-fleshed sort of Mr. Bakewell's, provided any able breeder, or body of breeders, would pay

---

* I never heard of any attempt to breed horses of any kind on this principle. If it has been attempted it must have failed. Although wild horses must be bred in-and-in as well as wild cattle, I have known instances of near blood-horse breeding by accident, and the results were certainly fatal to the principle.—*The Author.*

as much attention to these as Mr. Bakewell and his neighbours have done to the long-horns. But it has hitherto been the misfortune of the short-horn breeders to pursue the largest and biggest-boned ones as the best, without considering that those are the best that pay the most money for a given quantity of food. However, the ideas of our short-horn breeders being now more enlarged, and their minds more open to conviction, we may hope in a few years to see great improvements made in that breed of cattle."*

In the following pages I hope to point out how Mr. Bates, from his relation to and connection with Mr. Geo. Culley, became imbued with his spirit and maxims, and how he succeeded in realizing all his anticipations respecting the breeding of cattle and the improvement of the short-horn race. By the above quotations it is shown that at the time they were written the Messrs. Colling were not the sole, or even the most eminent, of short-horn breeders, or that their herds had acquired any remarkable value. The Messrs. Colling always said, however, that whatever they knew of breeding cattle they acquired from Geo. Culley.

The publication of the Herd Book, in 1822, in which Mr. Bates took an active part, was the next great event in short-horn history. George Coates, its author or editor, had been a successful breeder of short-horns, and a cow,

* To the edition of 1801 there is the following note:—"I am glad to find my hopes have been well founded; because, since the publication of the first edition of this work, a very rapid improvement has taken place in the breeding of short-horn cattle, so that in a few years I have reason to think they will surpass their rivals—the long-horns."

called the Driffield Cow, and bred by him, had been publicly exhibited, and a portrait of her published. His bull Patriot (486) was also considered a first-class animal. Mr. Coates, indeed, was well known, and a popular man with short-horn breeders. He, however, fell into reduced circumstances, and the publishing of a Short-horn Herd Book, after the model of the Racing Stud Book, was suggested as a means of providing for him. The share of Mr. Bates in this work will be referred to hereafter.

After Mr. R. Colling's sale in 1818, Mr. C. Mason, of Chilton, in the County of Durham, was considered the greatest breeder of short-horns. He was a gentleman well known and popular in the agricultural and sporting world.

Mr. Hutchinson, a banker at Stockton, had also a short-horn herd, and puffed off their praises in verse, attempting at the same time to cast ridicule on other herds in his doggerel rhyme.

Soon after this, Mr. Bates became acquainted with the Rev. Henry Berry,* and they often discussed the subject of breeding cattle. Mr. Bates, as usual, freely imparted to him all his knowledge and experience, and wished him to write the History of Short-horns, but Mr. Berry considered Mr. Bates the proper party to do so. Mr. Berry was an enthusiastic breeder of cattle, and was much impressed with the value of short-horns over every other breed. He also tried West Highland Cattle, and obtained some fine specimens from Northumberland. The difficulties and doubts, however, of the undertaking were considered very great. Mr. Berry had written much in praise of short-horns, and to show their superiority over

* See Farmer's Magazine, 1838, p. 5.

all other breeds. The intended work was never proceeded with. Mr. Berry however published much on breeding, and he considered that more was often due to the dam than to the sire, in the character of the produce, a conclusion, I believe, now admitted to be erroneous, at any rate in crosses by well improved short-horn bulls with short-horn cows, or any other breed of cattle.

After the Royal English Agricultural Society's first meeting at Oxford, in 1839, short-horns advanced greatly in favour as profit-giving cattle. Breeders and agriculturists in large numbers subsequently came to visit the herd of Mr. Bates, at Kirklevington, and with them Mr. Bates usually entered into the subject of short-horns and breeding in general. Mr. Bates stated so many facts respecting short-horns and their origin and history, and the defects of the various herds which had existed, that a very general wish was expressed to him, and frequent requests made, that he should commit his knowledge to writing, and in fact write a history of improved short-horn cattle. Mr. Bates frequently expressed his intention of doing so, when he had leisure for the purpose. He made notes and memoranda of short-horns, and the various anecdotes and events connected with them, and the owners of the herds which he had visited. The popular notion of improved short-horn cattle was, that the Messrs. Colling had possessed all the race, and been the real improvers of the breed. Nothing surprised the visitors more than the facts and personal observations of Mr. Bates, which extended as far back as the year 1782, respecting short-horns. Mr. Bates considered the Messrs. Colling neither the first breeders nor the improvers of short-horn cattle, and that, in fact, the celebrity of the Colling's herd and name, although made widely known by the travelling Durham Ox, calved in 1796, was caused by Mr. Bates purchasing the Duchess Cow, in 1804, for 100 guineas: before that they had never sold any cows or heifers at extraordinary prices. Mr. Bates's time was so much occupied with his herd, and in receiving and paying visits, generally in connexion with short-horns, that he made little progress with his history, and he never reduced it into any regular form. My connexion with Mr. Bates brought me into constant conversation with him, and I was familiar with all the transactions relating to his own herd, and was generally present both in viewing the cattle and at table when he was talking of short-horns. From these advantages, I had long been familiar with all the facts stated, and events referred to by Mr. Bates, and, indeed, he often asked me to write the intended history from his dictation, We however never commenced until the autumn of 1846.

In now committing this special knowledge to paper, I think it best, in the first place, to give Mr. Bates's own statements relating to breeding, as nearly as possible as I wrote them down. Secondly, to state such facts as have come to my knowledge, which corroborate or support the statements of Mr. Bates. And thirdly, to give some details to prove what the Kirklevington herd really was, and what it has been, and now is, that short-horn breeders may all see how far Mr. Bates's opinions and predictions have been justified by the experience of above twenty years since his death.

CHAPTER I.

PART I.

EARLY OBSERVATIONS, REFLECTIONS, AND MATURED VIEWS.

Mr. Bates, Mr. Wastell, of Burdon, and the Importance of Agricultural Improvement—Messrs. H. and W. Taylor, of Christon Bank, Opulent Northumberland Graziers—Extra Value of Improved Cattle in the Yorkshire and Durham Markets—Mr. Harrison at Barmpton—C. Colling's Bull, Favourite—Mr. Wastell Bred Mr. Jolly's Bull (337)—Mr. Maynard's Cow, Favourite, and the Duchesses, Daisys and Cherries—Mr. Wastell and an Authentic History of Mr. Jolly's Bull, Jolly—Mr. Jobson, Turvelaws, Wooler, and Another Romantic Incident in the life of Jolly—Mr. Wastell and his Influence on the Young Mind of Mr. Bates—Mr. Bakewell's Improved Leicesters, and a Black-faced Tup—How Animals Breed Back—Short-horns Great Milkers, Grazers, and Meat Producers—Various Yield of Milk per Diem—Mr. Wastell's Cow, Barforth, gave 36 Quarts—Mr. Bates on the Origin of Short-horns—Mr. Michael Dobinson, the Isle, Sedgefield, the Founder of the Best Tribes—The Durham Agricultural Society and its Influence on Breeding Short-horns—Mr. John Erasmus Blackett and the Studley Short-horns—Sir James Pennyman, his Tenantry, how he encouraged them, and the celebrated Studley Bull—Mr. Lakeland's, Mr. William Barker's, and Mr. John Brown's Bulls were by the Studley Bull—Sir Hugh Smithson, and the Tradition of the Stanwick Short-horns—Short-horn History to 1783.

Mr. Bates stated that, in 1782, his attention was drawn to the importance of agricultural improvements by Mr. Wastell, of Great Burdon, near Darlington, while on a visit in the southern part of the County of Northumberland. Mr. Wastell, was a man of superior information in his day. Mr. Bates always spoke of him as the first improver, and also as one of the best judges of short-horn cattle. From many farms in that district Mr. Wastell obtained, for many years in succession, the surplus stock for feeding, he furnishing the owners with bulls regularly as they needed them. He then came to buy the stock so descended, to graze upon his estate at Burdon. Mr. Wastell was a near relation of the Messrs Culley, and being a much older man, had been their adviser in their early years. He was also the intimate friend of Messrs. Henry and William Taylor, of Christon Bank, formerly the most opulent graziers in Northumberland, who accumulated an immense fortune from small beginnings, principally through their great skill and judgment in cattle. Mr. Wastell used to visit them to select their best cattle when nearly fat, that he might finish them at Burdon. The dealers who attended Wakefield Market, then the largest fat cattle market next to Smithfield in England, could give from sixpence to one shilling per stone, and sometimes even more than the butchers who bought at Morpeth and Darlington, who supplied the shipping at Shields, Newcastle, and Sunderland. Mr. Wastell was long recollected and spoken of as having the clearest conception and the most correct and sound judgment. His remarks were so indelibly impressed on the mind of Mr. Bates, that they were never obliterated. He often stated that as he grew up and gained increased experience, he invariably found that his, (Mr. Wastell's) judgment was correct. Pedigree was everything with Mr. Wastell; a long line of the best ancesters was indispensible, if men wished to breed to a certainty, not that he did not know what great judgment was requisite in putting the most proper males to suitable females. This, Mr. Wastell said, was the only way to make a permanent improvement in any breed of stock, always bearing in mind the purposes for which they are intended. He had for many years the very best short-horns in his possession. His brother-in-law, Mr. Harrison, resided at Barmpton before Mr. R. Colling began farming there.

In 1799, Mr. Bates twice bought steers descended from a cow bought at Mr. Harrison's sale after his death, and, although the times were very much depressed, this cow cost fifty guineas. Mr. Bates affirmed that these steers were better than any he ever saw, either at Barmpton or Ketton, when the Messrs. Colling's stock were at their greatest perfection. Mr. Bates had bought Mr. R. Colling's steers the following year, 1800, and both lots were by Mr. C. Colling's bull Favourite, then in his bloom.

It will surprise many, I doubt not, to hear that to Mr. Wastell the short-horn breeders are indebted for breeding Jolly's Bull (337), from which bull Mr. Maynard's cow, called Favourite, owed the excellences she possessed. These qualities, too, were transmitted to her descendants, even to the day of Comet, though crossed in the interim by bulls, not only of no celebrity, but with gross and glaring imperfections, which rendered that tribe of cattle far inferior to the Duchesses, Daisys, and Cherries of Mr. C. Colling's stock, and the Princess tribe of Mr. R. Colling. Mr. Bates stated he believed that the way Mr. Jolly's bull was bred was never made public till the summer of 1840, after the Yorkshire Agricultural Society's exhibition of that year, when Mr. Holmes, from Ireland, favoured Mr. Bates with his company at Kirklevington. He asked Mr. Bates if he knew where Mr. Thomas Robinson lived. He had been an eminent breeder of short-horns, and formerly lived at Marsh House, near Stockton. Mr. Bates said that Mr. Robinson then resided in the town of Yarm, next neighbour to Mr. Jolly, and as he (Mr. Holmes) wished to see Mr. T. Robinson, Mr. Bates asked him and Mr. Jolly to dinner. After dinner, Mr. Bates requested Mr. Jolly to inform Mr. Holmes where he got the bull named after himself, and what became of the bull afterwards. Mr. Jolly, then in his eighty-eighth year, but as hale and hearty as when he was half the age, said, "When I was seventeen years of age, having from childhood heard so much about the superiority of Mr. Wastell's cattle, and having seen Mr. Wastell frequently at Darlington market when I was there with my father, I felt an anxious desire to see them, and one day, on seeing Mr. Wastell getting his horse to leave the market, I went to him and asked him if he would kindly show me his cattle; to which he replied, "I am going home, young man, and if you will get your horse and accompany me I will show them to you." I accompanied him, and on arriving at Burdon, Mr. Wastell said, "I will take you to my cow-house at once;" and there seeing a young bull calf which I was much taken with (never having seen such a one before), I said "Will you sell me this calf?" and he said "Yes, if you will give thirty guineas for him;" to which I replied "Then I'll give you thirty guineas." He then said, "Young man, he is yours, and you may fetch him whenever you like." I went home and told my father that I had bought a young bull calf of Mr. Wastell for thirty guineas, and he was exceedingly angry with me for giving so high a price, and said I would never see the money again. But, as he

c

was intimate with Mr. Wastell, he gave me the thirty guineas the next morning, and told me to bring him home. I did so, and the bull was used for many years on the farm at Worsall.

Some years afterwards, while I was down in the Holmes, by the side of the River Tees, a gentleman came to me on horseback, and asked if I had the bull I had bought as a calf of Mr. Wastell, and I told him I had. He then asked if I would sell the bull, and I said I had no objection. We then returned to the farm-house, and I held his horse while he went and examined the bull, and when he came back he said, "What is the price of the bull?" I said, "fifty guineas," on which he took out his purse and told fifty guineas into my hand, and said he would send for him. His name was Mr. Jobson, from Turvelaws, near Wooler. This Mr. Jobson was father of Messrs. William and Robert Jobson. Their stock were descended, until the sale in 1846, from this same Jolly's bull, and though the purchase of Jolly's bull was made nearly seventy years before, they then retained the same character as was in Mr. Maynard's cow, Favourite. Mr. Maynard only sent one cow to this Jolly's bull, and she, far from being the best of cows at that day, yet produced the highest priced animal ever sold by public auction. Mr. Bates said he named all these facts particularly, that young breeders like young Mr. Jolly may not act as he did, for he never bought another good bull. Mr. Jolly attended Mr. Robert Colling's sale in 1818. The dam of Mr. Bates' second Hubback, was bought from Mr. Waldy, an agricultural pupil of Mr. Jolly. Mr. Waldy, senior, was very angry with his son for giving twenty-six pounds, twelve shillings, and the auctioneer's fee, for her, at Mr. Hustler's sale, at Aclam, although it was well known that Mr. Hustler had repeatedly refused 400 guineas for her. Mr. Waldy, as well as Mr. Jolly, had seen much worse cows sold at the Barmpton sale the previous year, at above 300 guineas each.

As I have mentioned, Mr. Wastell was the person who first aroused Mr. Bates's attention to the improvement of stock. Mr. Bates often mentioned a fact which showed Mr. Wastell's discernment, and it was this—"Mr. Bakewell," he said, "had concealed from the world how he had produced that wonderful improvement in his flock, but if the truth was known he was sure that it would prove he had done it by the use of a black tup." Seventeen years after Mr. Bates heard Mr. Wastell state this, he named it to a gentleman who had visited Mr. Bakewell in his early career as a breeder, and he said that "while staying with Mr. Bakewell I observed that there was one part of his premises he never showed me, and I got up very early one morning and went and examined those premises, and I there found a black tup; a most extraordinary sheep; but as it was a liberty I had taken, I never named it before to any one." This he told Mr. Bates, which confirmed what Mr. Wastell had said. Mr. Bates stated that he had learnt from good authority that Mr. Bakewell bought this black tup at Ashbourne market in Derbyshire, and that the fact was well known there by those who saw Mr. Bakewell buy the black tup. Mr. Bates stated that he had good authority for saying that of late years black lambs occasionally came from Leicester ewes, though none were ever seen in Mr. Bakewell's day, which is as strong a proof as can be given that *animals breed back* even to very remote ancestors, where there has been anything very peculiar in such ancestors; but the general principle holds good in most cases that "like begets like." Mr. Bates stated that breeding is a subject inexhaustable to an attentive observer; he may gain knowledge and experience each day he lives, and may ever keep improving.

Mr. Bates related many instances he had known of cows of the short-horned race that combined the character of great milkers, quick grazers, and of growing to immense weights. Mr. Bates remembered a cow belonging to Mr. Dixon, of Ingoe, Northumberland, that gave very large quantities of milk, often breeding calves that were all good grazers, was fed at the age of seventeen years, and made a very fat and handsome cow. These cattle were small consumers of hay. Mr. Bates mentioned these facts to Mr. Robert Colling when at Barmpton in 1804, before he bought his first Duchess cow of his brother Charles, and Mr. C. Colling said, "When I came to Barmpton first, after the death of Mr. Harrison, I never had then any thought of becoming a breeder of short-horns, and only kept cows for dairy purposes. My sister, who then lived with me, said that the milkmaid had repeatedly told her that one cow gave an extraordinary quantity of milk. This she told me one Sunday evening, before the cows were milked, and I said, 'Let us go and see what she does give, and we will measure the milk.'"

After seeing this cow milked, it measured twenty-six quarts and a-half, thus proving that she gave more than the cow to which I have before alluded. The next greatest quantity of milk Mr. Bates ever heard of a cow giving, was one which belonged to a person who said to him "She does more than fill the can, and the girl takes with her a smaller can," which, on measuring, was nineteen and a-half quarts regularly twice a day while on grass. The next largest quantity he remembered was eighteen quarts each end of the day, which was given by a cow belonging to Mr. Alexander Hall, and she was by Mr. Masterman's bull, whose descendants he afterwards sold to the Messrs. R. and C. Colling. The Bright Eyes family, afterwards called Princess, were bought by Mr. R. Colling. Mr. C. Colling's cow was called Houghton, and was the dam of the bull Foljamb. Mr. Wastell had a cow called Barforth that gave eighteen quarts of milk each meal, and made sixteen pounds of butter per week of twenty-four ounces to the pound. Mr. Bates never had a cow that gave (to his knowledge) more than fourteen quarts to a meal. His first Duchess, by Daisy Bull, gave that quantity, and each quart, when set up and churned separately, gave one and a-half ounces, or twenty-one ounces per meal. The butter was made up for the Newcastle market in half-pounds of ten and a-half ounces each, and was sold at one shilling per half-pound. The milk, after being creamed, was sold to the labourers at a penny per quart—which makes, at twenty-eight quarts per day, 16s. 4d. per week; and taking off 2s. for the diminution of the cream, and fourteen pounds of butter per week at 2s. per pound, making 28s.—this, added to the old milk value, makes better than two guineas per week. This she did for some time in the summer, having calved the 7th of June, 1807. She pastured with other nineteen cows, and was kept in the same way in every respect, getting no hand food whatever. Mr. Bates said, "This I think necessary to state, as many trials of milch cows have been made when they were getting hand food of a costly kind, such as Indian corn or linseed cake, which I never have given to milch cows." I also know that Mr. Robert Colling's Bright Eyes (the dam of Marske bull, which was sold at his sale) gave fifteen quarts of milk each time of milking. The late Mr. Hustler, of Aclam, who had Daisy Cow by Favourite, own sister to the Daisy Bull which I used, matched her as giving more milk when on "fog" in the autumn against his tenant's cow (Mr. Appleton's) who lived near him. On the milk being measured, Mr. Appleton's cow gave fifteen and a-half quarts, and the Daisy cow sixteen quarts. Twelve quarts of milk per meal is considered a good quantity when the cows only get ordinary keep. Match'em cow, the dam of the Oxford Premium cow, never gave less than twelve quarts per meal when on grass after calving."

On the origin of short-horns, Mr. Bates proceeded to say: "The question is often asked, whence did the short-horn cattle come, and how many years can they be traced back? I have heard it asserted that short-horn cows were originally sent from England to Holland, nearly two centuries ago, as a present by James the Second to William Prince of Orange, then Stadtholder, at the time of his marriage to his daughter, but those whom I have heard make the assertion, though they promised to furnish me with dates, have never done so. From this produce a century after, Sir William St. Quintin and others made their importations. But the best tribes were brought over by Mr. Michael Dobinson, who went in the early part of his life to Holland, in order to buy bulls, which statement is made by Mr. George Culley in his treatise on 'Live Stock.'" He says "those brought over were of much service in improving the breed, and this Mr. Dobinson and his neighbours, even in my day, were noted for having the best breed of short-horn cattle, and sold their bulls and heifers for very great prices." Now Mr. George Culley was born in the year 1730, and in or about this year, my grandfather, John Bates, of Aydon, near Hexham, being at Yarm, saw in the market a cow of this breed, belonging to Mr. Dobinson's brother, who lived at the Isle near Sedgefield. He went home with Mr. Dobinson and bought six cows and a white bull, from which stock he bred ever afterwards, and at his decease in 1777, they passed into the hands of my uncle, and I well remember the extraordinary good cattle they were up to the year 1800. I may here also state a fact not generally known, that after Mr. Robert Colling had bought in his breeding stock, my uncle went over to Barmpton, and purchased his best three cows, and the year following Mr. Robert Colling came and asked as a favour that he would let him have them again, which he did at prime cost. My uncle had kept them a year, and did not find them superior to his own cows.

The Agricultural Society for the County of Durham made a great demand for the best short-horns. Mr. Robert Colling said he foresaw this from its commencement. As the formation of this society created, as it were, a new era, I shall mention it more particularly hereafter, and proceed to give the earliest truly authentic account that I have been able to obtain of the best short-horn breed of cattle, which was given to me by the late John Erasmus Blackett, Esq., for a long period an alderman in the Newcastle corporation, and younger brother of the late Sir Edward Blackett, Bart., of Matfen, in Northumberland. He was born and brought up at Newby Hall, near Ripon, which was long the residence of that ancient family. It was, however, sold about a century ago, and is now the seat of Earl De Grey (late Lord Lieutenant of Ireland). I had frequent conversations with Mr. Blackett on the subject of short-horns, and he informed me that his great-grandfather had a valuable breed of short-horns, which were greatly improved by acting in conjunction with the family of Aislabie, of Studley Royal, a family property for above eight hundred years. These were both very ancient families, and might have had that breed at a much earlier date than Mr. Blackett spoke of. Mr. Blackett also said that the entrance hall at Newby was hung round with portraits of their most celebrated short-horns, taken by the best artists of the day. Mr. Blackett had a very strong retentive memory, and was a close observer of facts. When he visited me at Halton Castle, in 1804, after showing him my young cattle, bred from Mr. C. Colling's herd, I had two bulls brought out, the one Daisy bull, (186), and the other Styford (629), which I had hired at that time of Mr. R. Colling. The predecessors of both bulls were by Hubback (319), and showed his character in their looks, hair, and handling, in a very striking manner. After examining both bulls very attentively he said "they reminded him greatly of his father's cattle in his early years, and pointed to their distinguishing characters, from which I concluded they preserved the ancient character unimpaired. After the Blackett family left Newby, the Studley family continued breeding short-horns. It is a well ascertained fact that Sir William St. Quintin, of Scrampton, on the Yorkshire Wolds, and Sir James Pennyman, who had estates and residences both near Beverley and at Ormsby in Cleveland, acted in conjunction, and got their breeds of short-horns originally from Studley. Sir William Pennyman, now above eighty years of age, very kindly desired his steward, Mr. Rutter, to show me the farm accounts of the late Sir James Pennyman, which he most obligingly did. These farming accounts were kept by the agent, and go back far more than a century, in which are regularly recorded the sales of the cattle, their weights, and proof in tallow, &c., they being often sold by weight. As Ormsby is a strong clay soil no turnips were grown there, so that hay and straw only could be given the cattle in winter, yet the weights of the steers at four years old were generally about 140 stones of 14lbs. to the stone. There are also several entries of ten shillings and sixpence per head having been paid at different times for cows sent to the bull at Studley, proving that, not only the original tribe of short-horns was from Studley, but that they were repeatedly renewed when thought necessary and advantageous.

Sir James Pennyman, to induce his farm agents on each estate to exert themselves to pay attention to their respective stocks, frequently made small wagers as to whose oxen would make the greatest weights and prices. He was universally respected by all his farm agents, and no exertion was spared to do the best they could for their employer. We have, therefore, sure testimony in this tribe of stock to prove the high merits of the short-horns for above two centuries; and before Sir James ceased to breed cattle, his tenant (Mr. George Snowdon, of Hurworth) brought to his farm, in 1774, six cows and a bull from Sir James Pennyman's herd. One of these Mr. Robert Colling bought and called her Wildair, and she was the predecessor of Phenomenon, Wellington, &c. Two oxen, bred and fed by Sir Henry Grey, Bart., of Howick, in Northumberland, were by a bull of Sir James Pennyman's. Their weights are recorded by Mr. G. Culley in his treatise on short-horns:—"These oxen were killed in March, 1787, when seven years old. The red ox weighing 152 stones 9 pounds the four quarters; tallow, 16 stones 7 pounds; hide, 9 stones 2 pounds—178 stones 4 pounds, of 14lbs. to the stone. The mottled ox's four quarters weighed 152 stones 8 pounds; tallow, 16 stones; hide, 9 stones 11 pounds—178 stones 5 pounds."

Much about the same time with the above two eminent breeders of short-horns, the Milbank family, of Barmingham, also procured the Studley breed of short-horns, which they retained for a long period, as I have repeatedly been informed by the descendants of Mr. Milbank's steward, who have described to me the fine animals they produced. What attracted the attention of the steward's sons was the ribbons attached to the horns of the oxen when they were sent to be slaughtered. Mr. Sharter, one of this steward's family, commenced farming, as a tenant, Mr. Milbank's estate at Chilton (where the late Mr. C. Mason afterwards resided), and he took with him cows of Mr. Milbank's breed of short-horns, and a celebrated bull called the Studley bull.

An ox, five years old, bred and fed by Mr. Milbank of Barmingham, in Yorkshire, was killed at Barnard Castle in April, 1789, by Mr. Lonsdale, and his weight, as given by Mr. G. Culley, was "the four quarters 150 stones 4½lbs., tallow 16 stones, hide 10 stones 11lbs.—177 stones 1½lbs." From the above statement it appears that the Barmingham ox at five years old was of equal value with the Howick oxen at seven years old. Mr. Sharter, of Chilton, slaughtered a cow when twelve years old which had produced several calves and weighed upwards of 110 stones. This cow was a daughter of the old Studley bull (626). He is the ancestor of some of the most celebrated short-horns. I have heard him described by persons who had seen him as possessing a great girth and depth of fore-quarters, very short legs, a neat frame, and light offal. Mr. Lakeland's bull was a son of the Studley bull, and Mr. William Barker's bull (51), was a son of Lakeland's bull, and Mr. James Brown's old red bull (97), was bred by Mr. John Thompson of Golington Hall, and was by Mr. William Barker's bull, and he (Mr. J. Brown's old red bull) was one of the most celebrated bulls of his day. The first cow Mr. R. Colling bought of the old Red Rose tribe was by Mr. J. Brown's old red bull, and the cow bought at Stanwick, of the late Duke of Northumberland's agent, in 1784, was also by Mr. J. Brown's old red bull. Thus we have the regular pedigree of the short-horns brought down from the Studley stock.

The Smithsons, of Stanwick, as handed down by tradition from father to son for a long series of years, give an older date to the breed of short-horns in their possession than even that of Studley. Sir Hugh Smithson, the grandfather of the present Duke of Northumberland and also of Lord Prudhoe, in his early life, before his marriage with the heiress of the Percy family (now above a century ago), kept up the celebrity of this tribe of cattle. He was a great agriculturist and obtained the premium of the Society of Arts for planting, &c., &c. He paid the greatest attention to breeding and used regularly to weigh his cattle and the food they eat so as to ascertain the improvement made for the food consumed, and this was before Mr. Bakewell was known as a superior breeder of stock, and is the first authentic account known of this being done.

Lord Prudhoe very obligingly informed me that a very diligent search had been made amongst the old records at Stanwick, but they could not find any written documents.

Mr. C. Colling, who bought the cow at Stanwick, in 1784, has often described her to me as a very superior cow, particularly in her handling, of which no one was better able to judge. He told me repeatedly that he considered her the best cow he ever saw or ever had, and that he never could breed as good a one from her, though put to his best bulls that improved all his other cattle. Thus I have traced the ancient breeds of short-horns down to the period of the formation of the Durham Agricultural Society, which took place on the 15th of September, 1783, which was a year previous to the formation of the Highland Society in Scotland.

PART II.

THE PROGRESSIVE DEVELOPMENT OF THE REAL SHORT-HORNS.

The Durham Society's Prizes to 1797—Mr. C. Colling visited Mr. Bakewell—Mr. Bates on Mr. C. Colling's Judgment—The Cow DUCHESS—The Grey Bull Hubback—Mr. Bates, Mr. Hustler, Mr. C. Colling, and a Discussion on "Handling"—Styford, the Grey Bull Hubback and Belvedere (1706)—DUCHESS, a Stanwick Cow—Mr. C. Colling and Mr Booth—The Bull Hubback sold for One Guinea and 30 Guineas—Hubback, Mr. Bates's Description of—Hubback and the Herd Book—The Herd Book an Abortion: Erroneous Entries—The late Best Cow in England—Messrs. Colling acknowledge their Cardinal Error—Messrs. Culley's Herd—Hubback Cows and a Favourite Bull the source whence fine Short-horns sprung—Hair and Handling, or quality of flesh the only trustworthy indication of purity.

The Durham Agricultural Society was inaugurated in 1783. The grandfather of the present Duke of Cleveland, then Earl of Darlington, is the first name on the list of members. He was an eminent agriculturist and improver of both land and stock, and fed several fine oxen, which are mentioned by Arthur Young in the "Annals of Agriculture," and he continued to work and use one for draught long after they had been disused by others.

The premiums offered were for the best bull, five guineas, and for the second best, one guinea, the bulls to be shown at Durham, at St. Cuthbert's fair, the 31st March, 1784. For the best breeding cow in milk, three guineas, the cows to be shown at Darlington, September 20th, the same year. Mr. Christopher Hill, of Blackwell, received the premium for the best bull, and Mr. Joseph Robinson, of West Brandon, for the second best. Mr. C. Colling, of Skerningham, in the County of Durham, father of Messrs. R. and C. Colling, for the best breeding cow in milk, and Mr. R. Colling, of Barmpton, for the best tup.

In 1785, Mr. C. Colling obtained the premium for the best bull, afterwards named Hubback. Mr. Francis Walker, of Bradbury, for the second best bull.

In 1786, Mr. Francis Walker's bull obtained the highest premium, and Mr. C. Hill for the best cow.

In 1787, Mr. Thomas Hutchinson, of Sockburn, obtained the premium for the best bull, and Mr. R. Colling for the second best; Mr. C. Hill for the best cow, and Mr. R. Colling for the best heifer.

In 1788, Mr. R. Colling for the best bull, and Mr. C. Hill the second best; Mr. William Teesdale for the best cow, and Mr. C. Hill for the best heifer.

In 1789, Mr. John Burrell, of Mordon, for the best bull, and Mr. R. Colling for the best cow.

In 1790, Mr. C. Colling for the best bull, Mr. Joseph Grainger the second best, and Mr. R. Colling, the best heifer.

In 1791, Mr. G. Coates, of Haughton, for the best bull; and Mr. R. Colling for the best cow.

In 1792, Mr. C. Colling obtained the premium for the best bull (Lord Bolingbroke) and also for the best cow (Phœnix).

In 1796, Mr. C. Colling obtained the premium for the best bull (Favourite).

In 1797, Mr. M. Hutton, of Sledwick, for the best bull, and Mr. C. Colling for the second best. This bull was the sire of Mr. C. Colling's cow Lady, and this is the last time that the Messrs. Colling exhibited stock for premiums.

In the year 1783, Mr. C. Colling made a prolonged visit to Mr. Bakewell, at Dishley. The Christmas following he dined with his brother Robert. Mr. Wastell, of Ailey Hill, was also there; and after dinner Mr. Robt. Colling asked his brother if he knew of a bull that would do to serve his and Mr. Wastell's cows. He said they were then rearing a large bull calf, and only wanted one to use for a time, until this large calf was fit for use. Mr. R. Colling has frequently assured me that large cattle were then all he attended to or thought of, as he then knew of no other merit. Mr. Charles said he had seen a little bull in a field he went through to Church, and he thought he might be bought for very little, so that they might sell him without loss when their large calf was fit for service. He was then asked to buy the bull for them, which he did, the price being eight guineas. Mr. Charles Colling married soon afterwards, and he and his wife dined with Mr. Robert the Christmas following, and Mr. Wastell was again there. After dinner, Mr. Robert told his brother that he and Mr. Wastell had begun to use their large calf, and no longer required the services of the little bull he had bought for them, and asked if he could get them a customer for him. Mr. Charles asked what would be the price, and they both said they should be glad to take prime cost, on which Mr. Charles said they might send him to Ketton. However, some time passed over and the bull was not sent, and Mr. Charles began to think they had found out his value and intended keeping him. But having on one occasion to leave home for some time, on his return he found the bull had arrived, and he and his friends went out to see him, and he considered him "better than any bull he had ever seen." Soon afterwards Mr. Wastell sent a cow to him, and Mr. Charles told the servant who brought her that he might leave her and go home and tell his master he did not mean to serve cows with the bull under five guineas each, and that if he agreed to pay that price she might be bulled. He did so, and returned and said that his master would not pay five guineas for having a cow served by a bull for which he had only received four guineas as his half share.

Mr. Bates added: I have stated this thus particularly because a very different version has been given, and Mr. C. Colling has been reflected upon when there was no ground for complaint whatever, "except the want of judgment in the sellers of such a bull." Mr. C. Colling put his bull into condition, which he soon accomplished, as he was a most extraordinary quick feeder, and showed him at the Durham Agricultural Society's Show in 1785 (as before stated), where he attracted the attention of all real good judges, and proved the superior judgment of Mr. C. Colling in selecting him. He then began to buy up every good cow he could meet with. The cow from Stanwick he named DUCHESS. He bought many others from men who did not know their value, and parted with their best cattle at very low prices.

From this time a new era began in the breeding of short-horns, and *quality*, which had been long neglected, again became properly esteemed, as against mere size, as the criterion of merit. There had been a great depreciation in the value of farming stock at the conclusion of the American war, in 1782, and Ketton farm had made very little rent for some years. Mr. C. Colling's father was a great loser by that farm before his son Charles's marriage. Mr. Charles then began farming on the same farm on his own account, and from that time he selected good stock. No man ever had better fingers (so to speak) than Mr. C. Colling. Never had I to differ from him in opinion but once (though we often compared notes), and always till this instance did we agree. Mr. C. Colling was very confident he was right, and I was wrong. I remonstrated and said that I was right and he was wrong for once. The animals we differed in opinion about both belonged to Mr. R. Colling; the one was Styford which I had hired the previous year, the other was what is called the Grey bull, which Mr. Hustler and his tenant from Acklam had hired the year before. After a long discussion on handling, I asked Mr. C. Colling to go and re-examine both animals, and then say whether he continued in the same opinion or not. He did so, and then openly acknowledged his error, and said, "The Grey bull has precisely the same handling that Hubback had, and better than that of any other except the cow you bought of me yesterday evening, and her handling I consider the best, and all her predecessors have had the same handling." Mr. C. Colling always showed the Duchess family as the model of good handling.

I then asked the breeding of both these animals, and was told by Mr. R. Colling that they were from *two own sisters*. I then asked what bull they were each by? and the answer was Styford by Favourite, and the Grey bull by

D

the White bull. I enquired what bull got the dam of the White bull, and was told Hubback. And have you any cows now bred as the White bull? and he said, *none now in my possession.* Again, have you any bulls that are bred as the White bull, direct from Hubback to Favourite? and he said he had none. I then told Mr. R. Colling I had bought the only cow his brother then had, that went direct from Hubback to Favourite, and that I would give him his own price if he would let me send her to his White bull, but he would not at any price, and I offered him one hundred guineas to allow me to put her to him that they might see the result, which I said would be better than anything they ever saw; but though I repeated my offer, it was never complied with, and it was not till 1831 that I obtained the same blood in Belvedere (1706). No sooner had I bought him, than I told my acquaintances that I would produce, by the union of the Duchess and Princess blood, short-horns such as had never appeared before.

Mr. C. Colling bought the cow he called Duchess at Stanwick, and Mr. Alexander Hall's cow, the dam of Foljamb, soon after, and many others, but it was not till 1787 that he bought Mr. Maynard's cow, Favourite, and her daughter, the dam of Lord-Bolingbroke. But the most extraodinary part of this narrative is the fact that Mr. C. Colling, though so well aware of the merits of the bull afterwards called Hubback, yet in less than three years he actually sold him for thirty guineas, and for a long time offered him on sale before any one had the judgment to give him that sum. Amongst others who had the offer was the late Mr. Booth, father of the present Messrs. John and Richard Booth. Every calf by Hubback was like himself, however inferior the cows put to him, though of course the comparative value between those he did get was very great, owing to the difference in the dams. The same was the case with Belvedere. He never got a bad calf: but his calves, like those by Hubback, varied according to the dams. Those off second Hubback cows by Belvedere showed their superiority in every instance. I used Belvedere six years, from 1831 to 1837, and having his sons, Duke of Northumberland, then turned two years old, and Short Tail, three years old (both superior to their sire, although he was very good) I no longer had occasion to use him, and had him slaughtered. The dam of Hubback I never saw, but have heard her described by many who knew her well, that she was a small cow, as most good breeding cows are, her carcase was near the ground and was very fine in all points. She was a remarkably good handler when going in the lanes at Hurworth, and also an excellent milker, which was the object for which she was selected when her owner gave up farming and sold all his other stock.

Hubback and his dam were sold in Darlington market, in the year 1777, and the purchaser re-sold the calf for one guinea to a blacksmith of the name of Natrass, of Harrogate, near Darlington, at whose door he called in taking home the cow. The blacksmith gave the calf to his son-in-law, and it was brought up in the lanes at Hornby, within about eight miles of Kirklevington. Many of the tenants on the estate remembered this calf well, when going in the lanes at Hornby, and have described him to me most minutely as being admired by all who saw him. Two of the tenants were for many years the sole purchasers of short-horn cows for the London dairies. Hubback was afterwards sold to a person of the name of Fawcett, near Houghton, and re-sold before Mr. C. Colling bought him in 1783, at six years old. Hubback was one of the most remarkably quick feeders ever known, had clear waxy horns, mild bright eyes, and a very pleasing countenance. His handling was quite superior to any bull in his day; his coat was a soft downy hair, and he retained it long in the summer. His colour was yellow-red and white. His granddam was bought of Mr. Stephenson when he lived at Ketton, before Mr. C. Colling's day, and was got by Mr. Snowdon's bull. Mr. Stephenson, surgeon, son of Mr. Stephenson, of Ketton, described this cow as being from a tribe bred by his father for more than forty years. He left Ketton in 1769, and went to Hawick in Scotland. She was a small cow exceedingly neat and stylish, with particularly long and straight hind-quarters, was remarkably good in her hair, had a very great inclination to feed, and was a good milker, as were all the family of this tribe. She was sold at the Ketton sale on Mr. Stephenson's leaving, and was bought by Mr. Hunter, afterwards of Hurworth. Mr. Stephenson removed from Cleveland near to Ormsby, about 1731, and at that time the Pennymans had stock from Studley and were using bulls of that tribe. Mr. Snowdon's bull was bred on the dam's side from Sir James Pennyman's stock (who bred his cattle in conjunction with Sir William St. Quentin, as before mentioned), and on the sire's side by Mr. William Robson's bull, which was bred by Mr. Wastell, and the name of the dam of Mr. Robson's bull was Barforth, a most extraordinary cow in her day. I have often heard her described by many who had seen her at Mr. Wastell's. She was by Mr. James Masterman's bull (bred by Mr. Walker, near Leyburn), and Masterman's bull was by the old Studley bull. It was the opinion of all good judges in my early days that had it not been for the bull Hubback, and his descendants, the old valuable breed of short-horns would have been entirely lost, and that where Hubback's blood was wanting they had no real merit, and no stock ought to have been put in any herd book of short-horns which had not Hubback's blood in their veins. Had this been done, then the herd book of short-horns would have been a valuable record; as it is, it is undeserving of notice, and ought no longer to be continued as a book of reference, as ninty-nine animals out of a hundred in Coates's herd book should never have been entered there. Hubback was bought from Mr. C. Colling by a person of that name, living near Newbiggin, below Morpeth, in Northumberland, in 1787, when ten years old. This person never had a good bull before or since, but while in his possession he was used in the district, and all his descendants, even those from inferior cows, were much esteemed by breeders, all of whom admitted that they never had such stock as he produced. But none continued paying attention to their stock or reared bulls by him, and very soon no trace of his blood could be found in that district. The same was the case when he was used in Yorkshire, and in the County of Durham, from 1778 to 1784, till he came into Mr. C. Colling's possession, and then only three bulls by him were, I believe, kept for stock. One of these, named Broken Horn (95) in the Herd Book, is a wrong pedigree, as Mr. Colling never made two crossed by Hubback, and this Mr. C. Colling positively told Mr. Coates, when I accompanied him to Mr. C. Colling's, in order to get their pedigrees for the Herd Book. Again, in Punch (531) there are two crosses by Broken Horn inserted, which Mr. C. Colling told Mr. Coates was not the case; and, so far from being bred by Mr. George Best, of Manfield, was from an inferior cow belonging to Mr. C. Colling's father. When I read these statements in the first Herd Book, I named it to Mr. Coates the next time I saw him, and his answer was, "I had several good friends who had stock of their blood, and they wanted good pedigrees put to these bulls." It was not the fault of Mr. C. Colling that this was done, and Mr. R. Colling was just dead, and had no hand whatever in furnishing pedigrees, nor could Mr. Coates have given the old pedigrees in the first Herd Book, had I not accompanied him to Mr. C. Colling's often, when I took with me my written statements, which I had got from Messrs. R. and C. Colling many years before the Herd Book was thought of. In fact, till the preparation of the Herd Book was begun, very little attention had been paid to pedigrees except by myself; and those who had my pedigrees had got them from me, for the Messrs. Colling never refused to answer my various questions about the breeding and pedigrees of their stock. They, however, never gave me any information except when I asked them, and then they never refused. Unless, however, I had made out from close application, and knew all their stock and their descendants from observations repeatedly made, I could not have known it. But when I began to write the pedigrees down in their presence, and they thereby found I did know, they satisfied me fully by the information they gave.

The daughter of my first-bought Duchess I gave up in 1804, the day after buying her, at the urgent entreaties of Mr. C. Colling, as Mrs. Colling had so importuned him to request it, at the same time acknowledging I had bought her. Had I kept her I would not have needed to have bought the daughter at his sale in 1810, as I would then have had all the females of this family in my possession, and yet, he afterwards entreated me to let him keep the cow also, then in calf to Old Favourite, by my bull called Ketton, after the place he came from. In the meantime, he had refused performing the promise he made, that I should have her bulled by her son, then rising two years old, which was going into my neighbourhood, in Northumberland. This refusal was made, although the parties who had taken this bull had also agreed that she should be so bulled, and I told Mr. C. Colling that, had he kept his promise, I might have again given way to his entreaties as I had done in parting with her daughter, but that I would not do so after the treatment I had then met with. He said, "ask what sum you like to allow me to have her, and I will give it you." I told him, no sum would induce me then to part with her, on which he said "I acknowledge you are right; for you have in her the best cow in England." I told him I had long known that as well as he.

Mr. Bates then entered very fully into a statement as to the refusal of the use of the bull Duke, and his complaint of the conduct of Mr. C. Colling on the subject, and he continued by saying:—This was the only instance I ever had to find fault with Mr. C. Colling, and we afterwards continued on the best terms till his death. Mr. R. Colling died in 1820, and Mr. Charles in 1836. That the Messrs. Colling might have improved their cattle much more than they did, both brothers acknowledged to me; and that they should have used Hubback bull longer, and stuck more closely to his nearest blood they also acknowledged when they had parted with him. Their error arose from a too great desire to make money, rather than to improve the breed of short-horns. Both brothers repeatedly admitted to me that the short-horns were better before they came into their possession than they were even while they had them: this Mr. R. Colling repeated to me the very last time I saw him, in the autumn of 1819, when I stayed with him the greater part of two days, and went over the whole subject of short-horn breeding. He added, also, that the best lot of short-horns he ever saw together, in one herd, were those the Messrs. Culley took with them from Denton, near Darlington, into Northumberland, in 1767, when they entered on the Fenton farm, which was 17 years before he (Mr. R. Colling) began in earnest to buy up the best short-horns, and then it was not till after he had sold Hubback to his brother, admitting he *never saw his merits* till he had parted with him,* and then he determined to retain all he had got by him. What perhaps made the Messrs. Colling so unreserved with me was, they knew I had been present at many

---

* Mr. Carr, in his history of the Killerby and Warleby herds, mentions a similar case which occurred with Mr. Eastwood, who, having hired a bull, returned him at the expiration of a year. He was a small and rather shabby looking bull, but with great grazing propensities. He was fed off by the owner, and, just after his sale and delivery to the butcher, Mr. Eastwood drove over, and walking

meetings of older breeders than they, and who had known the short-horns in possession of Mr. Wastell. This was when they were, no doubt, at their very best, and about the time he sold Mr. Jolly the bull known as Jolly's bull, in 1770. This was before the American war which brought such distress upon the farming interest by the very low prices of agricultural produce, and from the money being drained out of the country to pay the expenses incurred in America. One of the memorable meetings of those who knew the best short-horns in early times, before Hubback's day, and who knew his descendants well, was in 1800, when, after viewing the Ketton Ox, then four years old, they all admitted that the best short-horns would be those that were bred from Hubback cows to a Favourite bull, and that all other crosses coming between had done the greatest harm. Of this remark I often reminded them, and they always admitted its truth. On this I have ever acted in my breeding of short-horns, and to this is owing, *entirely*, my success. Others would not see this, though I so often urged it, and I well remember a very shrewd man saying, when pressed upon the subject, "It is not that we do not know it, but it is not our interest to acknowledge it; you have that in your stock we cannot obtain, *hair and handling*, without which their can be no good short-horns; but those qualities are now gone, except in your herd."

round the yards, enquired for his shabby friend. "He is gone to the butcher," was the reply. "When? I have come to buy him," exclaimed Mr. Eastwood. "How could you think of sacrificing such a bull?" An express was sent off to arrest the axe, but unfortunately it was too late. It may be a question whether this propensity to graze was inherited from Hubback.

PART III.

THE DEVELOPMENT OF THE REAL SHORT-HORNS CONFIRMED.

Short-horn Breeders Discouraged—Imperfect Judges—Bolingbroke and a Red-polled Scotch Cow—The "Fool's Piece"—The Wonderful Ox—A more Wonderful Heifer—The Messrs. Colling's Great Service—Prophetic Vision, and Valuing a "Chance" at 2,000 Guineas—The Earl, the Issue of the "Chance" and his Premature Death—The Earl's Blood Preserved—Hubback's Female Issue and Belvedere the Foundation of the Bates Short-horns—Mr. Bates highly appreciates Duchess No. 1 in the Herd Book—The Yarborough Bull's Stock Good—A Confession by Mr. C. Colling, and Second Hubback's Handling—Major Rudd a Wrangler and no Judge—Hubback and Second Hubback the only True Ancestors of English Short-horns—Mr. Bates Despairs of Short-horn Breeders—Hereford and Devon Men better Judges—Mr. Mason's Confession—The Great Milking Qualities of the Bates Cows Demonstrated.

The Messrs R. and C. Colling had bought their best cows before the sale of Hubback in 1787, and Mr. C. Colling became discouraged in breeding by persons in that day, as is unfortunately the case even in the present day, preferring size, or large short-horns, instead of paying the proper attention to the quality of the flesh. This quality, however, is known only by those who really and truly can discriminate the same by handling. These imperfect judges then ran upon size and shapes, disregardful whether they are good handlers or not, and this caused Mr. C. Colling to sell Hubback before he had been three years in his possession, the first and most fatal error that he committed as a breeder. In 1789, and again in 1791, the Messrs. Colling sold short-horns to Mr. Robinson of Ladykirk, in Scotland, to Col., afterwards General Simpson, in Fifeshire, and sold Mr. Robinson at the same time the "Lame Bull," that was doing Mr. C. Colling the greatest harm in his herd, while he also sent cows to Mr. R. Colling's bulls that did both so much harm before they used the Favourite bull in 1795. For Mr. Charles Colling used the Lord Bolingbroke bull very little at that time, and Mr. Robert Colling not at all, as there was a shyness between the two brothers. Mrs. Colling's cow Phœnix, was sent to Ben, Mr. Robert Colling's bull, and the produce was Venus, and on being sent again to the same bull, after she had calved Venus, Mr. Robert Colling said to the servant who took her, "I wonder your mistress should send a cow to my bull," on which the servant perceived she was not to be bulled and took her home again. But, that she might breed a calf, on arriving home, she, Phœnix, was put to Lord Bolingbroke bull, and Favourite bull was the produce, and though so good a bull, Phœnix was never put to him again, but she was put to the grandson of Bolingbroke. This bull is No. 280 in the Herd Book, but he was a second cross from a Galloway red polled Scotch cow, the same year's calf with the Favourite bull; and the dam of which Galloway cross, by this polled Scotch tribe, was a hard-fleshed, bad handling tribe of short-horn cattle, quite the reverse of the Hubback blood, and though Lady of Phœnix was a fat rumped cow, and from giving very little milk and being well kept, was fat and sleek, she wanted hair and handling. All the stock, indeed, descended from her, wanted good hair and the proper handling, but being bad milkers they shewed fat on the tail-head, which is called the fool's piece. I never knew any of the tribe that were good milkers, and by using bulls of the same strain, the very best of short-horn cattle have been ruined. Wherever two crosses, on both sides, sire and dam may have been of this said blood, I never knew even a decent bull bred, however good-looking the cows were; and that has continued to be the case now for 50 years. They have deteriorated whatever herd they have been put to, though premiums are awarded to such at Agricultural Society's Meetings even in this day.

The first two calves by Favourite (252) were a heifer of the Duchess breed, and the ox (The Wonderful Ox) afterwards exhibited all over the kingdom, by Mr. Day, called the Durham or Ketton ox. These, when three years old, were shown at Darlington in the great market in 1799, and attracted a great crowd around them all the time they were in the market, from such animals never having been seen before so good at three years old. But the heifer of the Duchess tribe, was by far the best of the two, she was even more weight than the ox then was, and should have been kept as the show animal, but unfortunately Mr. C. Colling was advised to keep the ox and sell the heifer, which he did, and she weighed either 1lb. over 100 stones, of 14lb. to the stone, or 1lb. under, I do not recollect which, when only three years old. She was admitted by all who saw her to be the most perfect shaped and most uniformly covered carcase of beef, equally good in every point, with her breast the nearest to the ground of any animal ever seen. Her colour was a beautiful roan. Often has Mr. Colling said to me, "The heifer ought to have been the one kept for exhibition." The ox was kept on two years longer, and exhibited at Darlington on the first Monday in March, 1801, and was bought by Mr. Bulmer. Mr. Day's pamphlet also gives an account of the places he was shown at, and distances travelled, &c. His live weight was 270 stones. Mr. Charge's ox, which I was after being slaughtered, weighed 168st. 10lb., the four quarters, but having been unwell for some time before he was killed, had lost weight considerably.

Mr. Robert Colling's white heifer, when four years old, was estimated to weigh 130 stones, but she was not near so fine a heifer as the Duchess heifer at three years old, shewn with the ox at Darlington in 1799.

Mr. Arrowsmith's (Ferryhill) original short-horns were bought at Mr. Harrison's sale at Barmpton, in 1782. They were of the breed of Mr. Wastell, of Burdon, and at the lowest depression of prices towards the end of the American war, a young heifer cost 50 guineas. I bought Mr. Harrison's twin steers at Durham Fair, on March 31, 1799. They were by Favourite, and were extraordinarily good, and made a greater improvement for the time I had them than any I had then grazed; they, indeed, were even superior to those I grazed bought of Mr. R. Colling, October 19th, 1800, which were also got by Favourite, thus proving that Mr. Harrison's cattle were better than Mr. R. Colling's, when Messrs. Colling were at their very best.

The great service Messrs. Colling rendered to the country was in procuring the best short-horns then left, when they began breeding, for, had they not done this, the race would probably have been soon extinct. This was particularly the case as regards Mr. C. Colling's buying the Hubback Bull, and bringing him and his descendants into notice, for it was by Hubback's blood that the best of the old variety were preserved, and it is deeply to be regretted that Mr. C. Colling had not kept Hubback as long as he was serviceable. The beautiful white heifer he had from old Lady Maynard by Hubback, proved how well that cross answered, and yet he had only that one so bred, from having parted with Hubback. The introduction of Dicky Barker's blood, through Foljambe, was his next great error.

From the year 1806, when Favourite was slaughtered, the Messrs. Colling's herds became gradually worse, and though they continued selling at higher prices than before, no one derived any benefit from their herds. Of this I was convinced before the Ketton sale, in 1810. When my Ketton bull ceased to get calves in 1815, I then felt the loss, and I saw no way to reinstate myself in that blood but by sending Duchess cows to Duke (226), own brother to my Duchess I., in the Herd Book. When I had accomplished this I was convinced I could again keep improving. I remember in the summer of 1820, when Duchess 3rd was in calf to her uncle, naming this only hope of the short-horns to Lord Althorp, at Wiseton, when he said "I will give you 50 guineas for the chance, calf or no calf." I rejoined *I would not take two thousand guineas for the chance;* and I think any one who has seen my herd since must have been convinced that I was right in so highly estimating The Earl bull (646), then in his dam's womb.

I imprudently drove the bull calf, The Earl, when I removed from Halton Castle to Ridley Hall, May 1821, and turning him out to fresh grass on his arrival there after the day's journey, he swelled with the wet grass, being the first day he was turned out, and often afterwards did the same when on green food, clover, or turnips, and at last he was found dead at the stake, having had clover given him when the man went to church, and was dead on his return.

I had, however, kept three bull calves by him, one of which was my Second Hubback, the other two were named 2nd Earl and 3rd Earl. I had sons by Second Hubback which I used, and to the daughter of 2nd Hubback I put Belvedere. In this union of blood consists my present herd of short-horns, and they carry a character that no other short-horn blood possesses, and will transmit it to their posterity unimpaired, to the most remote generations, if judiciously put together. And in no other tribe of cattle can the best ancient blood of the short-horns be found pure and unadulterated.

From 1804 to 1810 I was less with the Messrs. Colling than at former periods, but as I had bought my Duchess, No. 1 in the Herd Book, at Mr. C. Colling's sale, I called some time afterwards, and Mrs. Colling said a gentleman at the sale had come to her after I had bought the Duchess heifer, and said, he "had asked Mr. Bates what he would take for his bargain?" and he (Mr. Bates) said he "would not take a thousand guineas for his bargain; was that so." I told her it was. She then said "how far would you have gone for her?" I said "I had not told anyone that," when she said "had I but known how far you would have gone you should have paid the uttermost farthing." I told her I well knew that, but I had taken the precaution to guard against it by not openly bidding myself, nor was it known till she was knocked down, when the auctioneer declared I was the purchaser at 183 guineas.

In 1807, on June 7th, my first Duchess Cow produced a heifer by St. John (572). I named her Baroness, and I had then two bulls from her, Ketton (709), and Laird (1158), and it is to Ketton that my cattle owe their superiority, for had I not possessed him I should not have bought Duchess 1st, in 1810. It was to put her to Ketton that I bought her, though I offered the buyer of Comet at the sale thirty guineas to have her put to him. My object was to try a second cross by Comet, though I was convinced she would breed better to Ketton than to any other bull that I knew of as being then in existence; Mr. R. Colling's White Bull not being then alive.

The old Daisy Cow, Mr. Hustler, of Acklam, then had (in 1807), and she bred his cow Fairy, after Mr. C. Colling's sale. The Acklam Red Rose Cow, own sister to Mr. R. Colling's No. 1 at his sale, and both own sisters to Styford (629), was returned to England from America, where she was for many years.

The Yarborough Bull (705), who came from his service at Acklam to the sale, was the cheapest and best bull bought there. Though only sold for fifty-five guineas he was better than Comet. Yarborough's stock invariably turned out well. The steers by him, in particular, had an uncommon growth, and astonished all who grazed them, for few if any were kept as bulls, he not being what was then considered handsome (usefullness then as now was not adequately regarded). All his female issue were also like his dam, very great milkers. He was long in the space between the "hook" and rib, which many descended from him show even now, though bred from cows called "close coupled," and some incline to what is called "baggy," a point found great fault with then and now, but not so by Mr. C. Colling. These two cows, Fairy and Acklam Red Rose, I bought after Mr. Hustler's death, in 1819. Soon afterwards I saw Mr. C. Colling in Darlington, when he said, "My brother Robert and I breakfasted with Mr. Hustler in going to the Murton exhibition, just before his death in that year, and we consulted together after seeing his cows. He said they, neither of them, ever bred so good a cow as the Acklam Red Rose, adding, she has exactly Hubback's handling." It was on this account I named her son Second Hubback, whose handling revived the impression the first Hubback made on the public mind, and those who knew the first Hubback agreed with me in considering the Second Hubback better than the first. There were several persons living at that time who had seen both bulls, and were good judges of cattle. Second Hubback's stock were all alike. I brought fifty females by him from Ridley Hall (where he was also calved), to Kirklevington, on May-day, 1830, all like beans, so nearly similar were they in every respect, in shape, colour, hair, and handling, as well as countenance, which never deceives a good judge of grazing cattle. All the cows by him were good milkers, without a single exception, even when put to Mr. Whitaker's cows, of Western Comet blood, which were small milkers, and of which he had two at his sale.

I may here mention the fact, that Norfolk's dam had two calves, sold at Mr. Whitaker's sale in 1833. Norfolk, by Second Hubback, sold for 124 guineas, and his half sister, by Frederick (267), sold for 31 guineas, just one-fourth the money. This showed the current of public opinion, notwithstanding the boasted excellence of Frederick's blood and produce, and the high condition of this heifer at the said sale. She was bought by Mr. Holmes to go to Ireland.

At Mr. C. Colling's sale in 1810, there was not one female I would have purchased, except the Duchess heifer, by Comet. Even the granddaughter of the excellent Daisy cow, by Favourite, that sold for 410 guineas, the highest priced female at the sale, was a poor creature, and I told Mr. Hustler at the sale that the granddam, even if she bred only one calf, was worth more money than this 410 guinea cow. Three years afterwards, when dining with Major Rudd, in Cleveland, I said to him privately,—"I do not know how your 410 guinea cow has turned out, but I would not have given you 20 guineas for her at the sale after you had bought her." His answer was,—"She never was worth 20 guineas to me, for she only bred one calf, a bull, and it could get no calves, and therefore was of no use." Thus I had not judged wrongly, and yet she was by Comet. Major Rudd was a wrangler, and often boasted of his high honors at Cambridge, but it did not show itself in breeding short-horns, as his pocket afterwards witnessed against him. He was only one of many, however, who have lost much money by breeding short-horns, which are not profitable, except when bred with judgment, and then they are and have always been so. But few, very few indeed, know the good ones, and judgment in short-horns, as exhibited in the decision at agricultural meetings, is worse than ever. I knew this at any period of my life from 1782 to this hour, though there are some good judges even in this day. I hope for the breed's sake and the benefit of the country that the number may increase, and that men may become wiser as well as better for their own sakes; but I say without hesitation, that if I had to begin breeding I would not buy a short-horn that had not Hubback or Second Hubback's blood, let their form and shape be ever so good, for without the grand requisites of good hair, handling, and style, short-horns never were and never can be good animals.

I almost despair of seeing short-horn breeders learning to know the good ones. The Hereford and Devon breeders know the best short-horns, better than the breeders of short-horns in general. Mr. Mason once said to me, "you can go on breeding short-horns, because they pay you in milk, butter, and beef, but we cannot unless we can sell at high prices to breeders." This confession was unguardedly made one morning when he called on me to breakfast, just as my housekeeper had put the week's butter in readiness for the Newcastle Market on the Saturday. I told him that, however ready he was for breakfast, he should not have it until he had counted the butter. There were three hundred half-pounds to go to market, besides what was sold at home and used in the house. There were then, I remember, thirty cows which had calved, and the butter sold for above one shilling per half-pound, being above ten shillings per cow in butter alone, besides the value of the old milk otherwise sold; while all the calves were reared by the pail, as I did not allow any calves to suck the cows. Had all the milk been creamed and made into butter, it would have been above twice the quantity. He however at that time, as I told him, kept three lots of cows, one to breed calves and then get dry, which was no hard matter, to attract notice by their high condition; a second lot, as wet-nurses to rear the calves; and a third lot, to supply the family with milk and butter. This is a system that would ruin any man, if he had the land rent free and no outgoings to pay, and yet many, even in the present day, pursue this reckless course to gain premiums, attract public attention, and gratify their vanity at the cost of the pocket.

PART IV.

THE PEDIGREES OF THE REAL SHORT-HORNS GIVEN IN FULL.

Hubback bought by Mr. C. Colling in 1785—To what Mr. Bates's success is entirely due—Mr. C. Colling committed another Great Error—Dicky Barker's Bull Described—Lady Maynard not a True Breeder—Bolingbroke's Issue not Uniform—Favourite's Issue a trifle Coarse—Authentic Documentary History of the Pure Short-horns: Mr. Alexander Hall's Hand and Seal—Dicky Barker's Bull—Mr. C. Colling Bought his First Duchess Cow for £13 in 1784—Lord Bolingbroke Sold for 70 Guineas in 1797—Johanna, the Lame Bull, Styford, Cupid, Yarborough, Venus, North Star, Comet, Stick-a-Bitch, Dolly, Princess (by Hubback), Princess Heifer, Foljamb, Daisy, and several other Notable Animals (Male and Female) Named and Described, and their Owners and the Date of their Sales and other Particulars Recorded.

In 1785, Mr. Charles Colling came into possession of Hubback, and had also the best females then in existence. Mr. Robert Colling having bought the cow from Mr. Hall, also his cow of the Red Rose tribe from Mr. Watson, of Manfield, and likewise Mr. George Snowdon's cow, put them to Hubback, and the descendants of these three were the only cows which possessed Hubback blood of all that were sold at Mr. R. Colling's sale, in 1818. This he told me before the sale, and he also stated the same the following year, the last time I saw him. Mr. C. Colling's cow of the Daisy family was by Hubback, and he always told me he had no doubt she had also a cross of Masterman's Bull (422 and 670), and this tribe was second to the Duchess family for being good milkers. The Cherry tribe of cows, which he got from his father's herd, and this tribe, through the dam of Yarborough, went direct from Hubback to his bull Favourite, and she and my first Duchess were the only two cows he ever had that were so bred. He had no bull so bred, so that we have here in these two tribes, Duchess and Cherry, and in Mr. R. Colling's stock, the Princess heifer (sold to Sir Henry Vane Tempest, of Wynyard), the only cow he had so bred. The only bull so bred was his White Bull (151), which he let to Mr. Smeaton for several seasons, and whose best stock went into the hands of his son-in-law, Mr. Edwards, of Market Weighton, at whose sale at Castle Howard, in 1839, I bought the Foggathorpe cow for £113, with auctioneer's fee. This animal I put to my Duke of Northumberland, from which, by the Duke of Northumberland, I have now two cows, and their descendants are very promising. Thus my breed of short-horns have at the present day the only blood so descended direct from Hubback to Favourite. To this blood I have kept throughout my breed of short-horns, and wherever I have in the least deviated therefrom, I immediately saw my error and parted with the produce, not breeding from them again, and it was by keeping to this rule *that my success has been entirely due.* By no other means could I have preserved my present herd in their purity, and I now have them as smaller consumers of food, and with the capacity of greater improvement on that less consumption of food than I have ever seen possessed by short-horns, even in their best days. I can have no doubt of the excellence of the ancient blood of short-horns before Messrs. Colling's day, but from all the accounts I could ever collect, they kept a less quantity of cattle per acre in those days, and consumed a greater quantity of food, although they got to great weights. Another great error Mr. C. Colling committed was the sending of his neat fine cow by Hubback to Mr. Richard Barker's bull (52), as his character was the very reverse of hers. I have repeatedly heard Dicky Barker's bull described by those who remember him well, and they all gave the same account which corresponds with that of Col. Trotter, of Staindrop—one of the purchasers of Comet—who, in a letter to me, dated June 17th, 1846, says "The bull Dicky Barker, was a large, coarse, wire haired beast, was by Mr. Hill's bull, and his dam a big coarse cow, with large head, dark horns, and black nose, his colour dark red flecked." This agrees precisely with what old Trotter said at his tup show, at Houghton, above forty years ago, when there was a large company of short-horn breeders who heard it, and they could never forget it, from the circumstance that caused this statement to be made. It was heard too with no small surprise by some who were just beginning to breed short-horns, for they considered them to have no drawbacks till they heard this statement, which was then all new to them, and confounded them not a little: the closer their enquiries, too, the more were brought out the stronger facts of the case. These no one could contradict who ever saw the Dicky Barker bull.

The produce of this neat cow was named Foljamb (263) after the person who bought him. Mr. Colling said he was not of good quality and could not be made fat. Foljamb was sold the same autumn that Hubback was sold, 1787, and only served a very few cows. Mrs. Colling's cow, Phœnix, was one of the produce and took the character of her sire, which was the very reverse to her dam, the bautiful Lady Maynard. I remember having a large party at Halton Castle in July, in the year 1812, and Mr. Mason, of Chilton, and Mr. Thompson, late of Stamford, in Northumberland, were present, when the breed of short-horns became the topic of conversation, on which I appealed to Mr. Thompson, who remembered well both Lady Maynard and Phœnix, if he ever knew one descended from Lady Maynard equal to herself, and he immediately answered, "None from her were equal to herself." I asked Mr. Mason if he could say anything to the contrary, and he was silent, as were the whole of the party; therefore in addition to my own view, here was the opinion of one much older, and whose judgment of short-horns and Leicester sheep was never called in question by any one. Lord Bolingbroke (86) was also by Foljamb, and from the daughter of Lady Maynard, but quite an inferior cow, with much coarseness, she having taken after her sire, Dalton Duke (188). Yet Lord Bolingbroke took to the character of Hubback, being cleanly and neat, inclinable to be podgy or big bellied and rather low sided. His stock were not one like another, but unequal. His son, Favourite, was calved in 1793, was from Phœnix, and partook more of her character. He was a large massive animal, had remarkably good loins and long level hind quarters; his shoulder points stood wide, and were somewhat coarse and stood forward into the neck ; his horns also were long and strong. These qualities he had from Mr. Hill's stock, the sire of Dicky Barker's bull, and all descended from Favourite, partook more or less of his shoulders. He was a powerful animal and of good constitution.

The bull calves by Favourite were generally like himself, a trifle course, but of good constitutions. Those which were better and got better stock than himself, were those more closely allied to Hubback, and were Mr. R. Colling's White bull (151), Ketton (709), Duke (224), Daisy (186), and Styford (629), all of which exceeded their sire, and had the Messrs. Colling used these bulls more freely, their stock, instead of going back as they did, would have greatly improved up to their sales. There are now none left of these once valuable tribe with uncorrupted pedigrees, except those in my possession. The public exhibitions since 1838 have shown none of the old valuable character except my own.

As many may be desirous of knowing every particular of the old variety of short-horns with which the Messrs. Colling began breeding, and as there can be now no good short-horns that have not their blood, I shall give many particulars from what I wrote down from the mouths of Messrs. Colling and of those who were contemporaries with them when they began breeding. I shall state a few of those particulars that future ages may know what the short-horns then were. Mr. Alexander Hall made the following statement to me on March 22nd, 1820, when I had not any of the Princess family, and was collecting information to supply Mr. George Coates before he began writing the first Herd Book. This document I have under Mr. Hall's own hand, and what I state I do from written documents, with the parties names to them who gave me the information. Mr. Hall says—"Mr. C. Colling bought of me a heifer by Mr. William Fawcett's bull, afterwards called Hubback, which heifer bred the bull Mr. George Coates bought of Mr. C. Colling for Mr. Foljamb. This heifer's dam was by a bull of Mr. C. Colling's, senior, and was bought by him of Mr. John Bamlet, senior, of Norton. Her grandam was by Mr. Harrison's bull of Barmpton, bred out of Mr. Wastell's old cow that bred Mr. William Robson's bull Foljamb (the grandsire of Hubback), and was by Mr. Richard Barker's bull from a cow of his (Mr. Barker's) own, and by Mr. Hill's Red Bull, which was by a half-brother of the Dalton Bull, and from a cow of Mr. Hill's own that was by Mr. Robson's bull. This cow's dam was greatly admired by the present Earl of Darlington's father (grandfather to the present Duke of Cleveland). The late Mr. R. Colling bought a twin heifer of me that was by Mr. Snowdon's bull (the sire of Hubback), and he bought her sister a year or two afterwards; their dam was by Mr. James Masterman's bull (great-grandsire of Hubback), and from Mr. Thomas Hall's great cow which was sold to the Duchess of Athol. This last cow was by Mr. Harrison's bull, that was bred by Mr. Wastell, from the same cow with Mr. Robson's bull, and from that famous cow of Mr. Thomas Hall's which was so great a grazer: Mr. Wastell called her Tripes. The dam of the twins, sold to the late Mr. Robert Colling, when getting only grass, gave eighteen quarts of milk (ale measure), twice a day, for more than six weeks together, in June and July; this we know, for, as we sold our milk at Darlington twice a day, it was measured regularly. The late Mr. R. Colling called the first twin heifer bought from me Bright Eyes, from her having remarkably bright eyes. The predecessors of this Bright Eyes, and the predecessors of Foljamb's dam, both by Mr. Harrison's bull, were from own sisters, daughters of Tripes. My great cow, from the dam of Bright Eyes, was put dry in October, and sold to the butcher the first Monday in March, at twenty-five guineas, and weighed 84 stones. Mr. John Hunter, of Hurworth, bred Hubback and his dam, which was by a bull of Mr. Banks's, of Hurworth, which had a great belly, but he was out of a handsome cow of Mr. Banks's own. The grandam of Hubback Mr. Hunter bought of Mr. Stephenson, of Ketton, who rented that farm before Mr. C. Colling took it. Mr. Thomas Hall (brother of Mr. Alexander Hall), bought the cow which Mr. Wastell called Tripes of Mr. C. Pickering, of Foxton, near Sedgefield, grandfather of Mr. Christopher Mason, of Chilton, but how she was bred it was not known. The Dalton bull was bred by Mr. Robert Charge, of Low Fields, brother to Mr. John Charge, of Newton, and was by a Red Hazel Bull of Mr. John Charge's, of Newton. This Hazel Bull was bred by Mr. William Dobson, of Croft, and by Mr. Duke Wetherall's old Red Bull. Mr. Thomas Hall's heifer, sister to Bright Eyes' dam, was sold to go to the south for twenty-five guineas. Another half-sister to Bright Eyes' dam, by old Mr. C. Colling's bull, was sold to Mr. Hill, of Blackwell, for twenty-five guineas; she bred a heifer that was matched against one of Mr. Hammond's, of Hutton Bonville, and won the match. Mr. Hammond's stock in that day were in great repute. A bull, a year old, from the great cow, daughter of Tripes, was sold for twenty-six guineas to Messrs. Bell, of Halton, in Northumberland, and was also by old Mr. C. Colling's bull. The stock from Masterman's Bull were all with waxy horns, but Hubback's stock were the best of any I ever bred. Their coats were like a moleskin, and their hair stayed on till midsummer. I am quite clear in the whole of the above statements, and am sure there are perfectly correct. Mr. Snowdon, of Hurworth, told me his stock were from Sir William St. Quintin's herd. His cow that bred Hubback's sire was a very handsome one, and remarkable for her wide 'hooks' and fine quick eyes; but the bull, sire of Hubback, drooped in his hind quarters, and was low sided; but he had a fine fore-end and good crops. These particulars I well remember.—Signed by me in the 66th year of my age, ALEXANDER HALL.—March 22nd, 1820."

I see in a memorandum I then made that he (Mr. Hall) was six years old when his brother brought home the cow which Mr. Wastell called Tripes, so that is bringing it to the year 1760. New milk was then sold 3 gills for ½d., now 1½d., three times that price. But in my younger days I knew several old persons in the same district, much older than Mr. Alexander Hall, who gave me the same statements Mr. Hall has made above; one of which is that the sire of Mr. William Dobson's bull, of Croft, was bought of Mr. Jacob Smith, near Ripon. That Dicky Barker's bull was a big beast, could not be made fat, but was well shaped; his flesh, however, was not of a good quality. His sire, Mr. Hill's bull, was sold to go to Ireland.

That Mr. Wastell's bull, which was given to his nephew, Mr. Robson, was by Masterman's bull. I showed these statements, as made to me, to Mr. C. Colling some years afterwards, and he said they were all correct and, he added, "Many of the best short-horns are low sided and baggy, particularly those that have the finest flesh and are the best handlers, and to this day this is frequently the case, but without the proper handling it is of no use having good shapes." In this opinion I coincided entirely with Mr. C. Colling when speaking of the character of the best short-horns.

The cow Mr. C. Colling bought of Mr. Maynard, Sept. 30th, 1786, was then seven years old and her price was £30 and twelve shillings returned, the price of her calf, to be taken the next spring following, was to be ten guineas.

I saw the above cow sold in Darlington Market on March 5th, 1798. There were two others, young ones, sold with her at twenty guineas each, but the old cow, though nineteen years old, and had bred twenty calves in all, was still a living fresh looking cow, and certainly was a far better cow than any bred from her that grew up to maturity. Masterman's bull's horns were a waxy colour and he had yellowish red spots all over him, but he was mostly white.

A white two year old heifer from Mr. Maynard's old cow Favourite, by Hubback, had an accident and died; she was much admired by Mr. C. Colling, as exceeding the dam, and proves how wrong he was in parting with Hubback. On June 14th, 1784, Mr. C. Colling bought his first Duchess cow, price thirteen pounds, a massive, short-legged cow, breast near the ground, a great grower, with wide back, and of a beautiful yellowish red flaked colour.

On January 9th, 1797, Mr. Wm. Jobling, of Newton Hall, Northumberland, bought Lord Bolingbroke at 70 guineas. Young Bolingbroke was let to Mr. Clover at 60 guineas. Mr. Charge in 1799, only paid 30 guineas a season for Favourite bull, and Mr. R. Colling and Mr. Mason only paid 20 guineas per year for the use of Favourite from 1800, no other person being allowed to send cows except Mr. C. Colling.

Johanna, lot 6, at Mr. C. Colling's sale, was by Favourite, from Johanna, by the Lame bull, and her grandam, bought of Mr. Finfield, of Chatham. Mr. C. Colling said he believed she was a sister by the same bull (the Lame bull, sold to Mr. Robinson of Ladykirk, 1791) as Colonel's dam sold to Col. Simpson. Colonel was bought by Mr. Thos. Jobling, of Styford, in Tyneside, and used some years by him, and was slaughtered before Thomas Jobling took Styford bull of Mr. R. Colling in the autumn of 1799. Styford returning to Barmpton, October 1800, when Daisy bull went from me at Halton to Styford and I bought him (the Daisy bull) of Mr. Thos. Jobling for 30 guineas in August 1802, and used him till the spring of 1804, when I took Styford of Mr. R. Colling.

Cupid, the sire of Yarborough, was by a son of Favourite, and calved at Christmas in 1799. General Simpson bought a daughter of Venus in 1806, which went to him that autumn. He also bought North Star, but did not take him till the Candlemas following, being then a year old. The sister of North Star came between.

Comet was calved in 1804 and North Star in 1806. On Friday, Sept. 11th, 1795, the first calf by old Favourite was calved by the Duchess cow. On Wednesday, October 7th, 1795, the calf which grew to be the fat ox was calved by a cow from the Stick-a-Bitch cow, the second issue of old Favourite. Dolly in 1800 was put to Cupid. A year old Cherry and a three years old Cherry, were also in 1800 both put to Cupid, from which Yarborough came. Miss Allen's three years old steers, by Mr. James Brown's old Red bull, weighed each 94 stones.

Dolly was put to Daisy Bull in 1799, and calved Wednesday, 8th April, 1800. Washington, by Daisy, dam Lady, was calved in 1800. Thursday, 12th March, 1800, my 1st Duchess calved. In 1796 Lord Carlisle's bull was calved. Mr. Mason's Lilly was by Favourite in the same year.

Dolly, by Punch, Mr. C. Colling says he could trace no further back. Dolly's daughter Mr. Hustler bought at the price of 50 guineas.

Yarborough's dam, by Favourite, her dam by Hubback from old Cherry, was a very mellow soft handler, and was a very great favourite with Mr. C. Colling.

In 1789, General Watson, of Fifeshire (Abadour), bought Princess by Hubback.

In October, 1789, Mr. C. Colling sold to Mr. George Culley a bull from Princess heifer.

The sire of Cupid, son of Favourite, was by Favourite, and from a Cherry cow. Old Phœnix had a heifer calf by this sire of Cupid, son of Favourite; and a daughter of her (by Mr. Bates's first bull) bred Mr. Baker's Surplice, by Favourite. He had very low flat sides, and was an excessively hard handler.

The Lame bull was exchanged for a bull by Hubback in 1788, and seven guineas were given in the bargain, on June 7th of that year. Mr. Robinson bought the Lame bull in April, 1791.

In September, 1803, Mr. Frost took Windsor, and bought a heifer by Favourite from Johanna. The first Cherry was bought at Yarm Fair by the Messrs. Colling's father.

On December 15th, 1793, the old Favourite bull was calved, and he died October, 1806, then in his 13th year.

In 1796 or '98 sold Mr. Compton a Johanna cow. Mr. C. Colling bought at Brafferton, a mile from Ketton, his first Daisy cow, an animal which was very inclinable to make fat, and very neat in shape. In 1804, in the spring, Comet was calved, and died June, 1815.

On December 14, 1787, Mr. C. Colling sold his bull Foljamb to Mr. G. Coates, for Mr. Foljamb. Mr. G. Coates observed that Hubback's quarters were long and straight, great breast, flank, and twist, pens rather down and fat, not formed to the hooks from the rump; coat soft, but not very long.

CHAPTER II.

PART I.

WHAT MR. BATES CONSIDERED A REAL SHORT-HORN, WITH AMPLE COLLATERAL EVIDENCE IN SUPPORT OF HIS JUDGMENT.

Pure Short-horns—Mr. George Culley on Purity and Quality as instinctively appreciated by Handling—Mr. Culley and Mr. Bakewell—Mr. Bates on Decisions at Agricultural Shows; his Success at York in 1838—Short-horns should be Shown in Families—Belvedere, Waterloo, Angelina 2nd, and Young Wynyard—Mr. C. Colling and Sir H. V. Tempest, Bart., and Princess and Wellington—More Pedigrees in detail, and Young Wynyard in Ireland—Authentic Documentary Proofs—Twin Heifers Breeding—Judges at Shows overlook Milking Qualities and heed only Fatness—Locomotive went to Kentucky—The Oxford Premium Cow bred Duke of Wellington, which went to Troy, New York—Heavy Weights of Pure Short-horns accounted for—Pure Short-horns will Improve from Generation to Generation—Mr. Bates begun Weighing Food and Measuring Cattle, which practice he recommends to all Breeders—Tyneside Short-horns Improved between 1785 to 1795.

It may be asked,—What are the distinguising characters of the best short-horns? The eyes are much more prominent than those of other cattle, the head much smaller, the tail lighter and finer, the skin and hair all over much softer, and the touch incomparably more kind and mellow. But the difficulty is to convey in language that idea or the sensation we acquire by the touch, or feel of the fingers, which enables us to form a judgment when we are handling an animal intended to be fatted. Mr. George Culley who felt that difficulty thus wrote :—" A nice or good judge of cattle and sheep, with a slight touch of the fingers upon the fatting points of the animal, viz., the hips, rumps, ribs, flank, breast, twist, shoulder score, &c., will know immediately whether it will make fat or not, and in which part it will be the fattest. It is very easy to know where an animal is fattest which is already made fat, because we can evidently feel a substance or quantity of fat upon all those parts which are denominated the fatty points; but the difficulty is to explain how we know or distinguish animals in a lean state, which will make fat, and which will not, or rather, which will make fat in such and such points or parts, and not in others; which a person of judgment *(in practice)* can tell, as it were, instantaneously: I say *in practice*, because I believe that the best of judges *out of practice* are not able to judge with precision, at least I am not. We say this beast *touches* nicely upon its ribs, hips, &c., &c., because we find a mellow pleasant feel on those parts; but we do not say soft; because there are some of this same sort of animals which have a soft loose handle of which we do not approve, and, though soft and loose, they have not that mellow feel above mentioned; for, though they handle both loose and soft, yet we know that the one will make fat, and that the other will not; and in this lies the difficulty of the explanation. We clearly find a particular kindness or pleasantness in the feel of the one, much superior to the other, by which we immediately conclude that this will make fat, and the other not so fat; and in this a person of judgment, and *in practice*, is very seldom mistaken. I shall only make one more remark, which is, that though the one animal will make remarkably fat, and the other will scarcely improve at all, with the same keeping, yet between these extremes are numberless gradations, which the complete judge can distinguish with wonderful precision."

I have made this long quotation from Mr. Culley, as he had been most extensively engaged as a breeder and feeder of stock for a very long period, and was in his 65th year when this was printed (1794), and it was above thirty years after he had spent one whole year with Mr. Bakewell, at Dishley, where he went in 1762, and was then above thirty years of age, and had been constantly engaged in farming before he went to Mr. Bakewell. I also chose to quote the words of so old a breeder at so remote a period, knowing well that in this age and country there are so *very, very,* few breeders,—and particularly the short-horn breeders—or graziers, that can distinguish by handling the quality of an animal. Mr. Bates then gave numerous instances of decisions of the most extraordinary kind by judges at shows, and he said that "in fact, to estimate their judgment by their decisions at agricultural meetings, I am compelled to say they either do not know a good short-horn, or else they give their decisions directly contrary to their convictions; and that animals having no claim whatever to breeding obtained the highest premiums, and were never heard of again." Mr. Bates felt convinced that the judges were appointed, not from their knowledge of cattle, but from their position, and because they could make what he called "clap-trap after-dinner speeches" on all occasions.

Although Mr. Bates got five premiums at York, in 1838, he always maintained that the best three short-horns were passed unnoticed. Mr. Bates said, "at the request of many breeders, I had the seven animals I had exhibited at York shown together in the yard when the stock got leave to be removed from where they stood in their respective classes, and this gave a more general satisfaction, as they then showed they were all of one family, corresponding in all respects. This is the only true way of exhibiting any breed of animals, and yet it is never done.* And why? Because no other breed of short-horns shew a family likeness except my own. Nor has any other breed of short-horns the same hair and handling that mine have; nor can it be obtained but through my strain of blood: for it runs in the blood, and none now can be found that have the old Hubback blood, and that of his predecessors, and of Mr. James Brown's old Red Bull, and these two bulls were the last remains of those breeds which had been so long eminent as short-horns, and it is now nearly sixty years since their day; and the proportion of this blood, after so many crosses with cattle of a contrary stamp, cannot be brought back to what they once were except by again resorting to the stock from whence that high merit came, which was before the time when the Messrs. Colling began breeding short-horns.

In 1831, I bought Belvedere from Mr. Stephenson, who bred him and his sire Waterloo (2816), and dam Angelina II. own brother and sister, both by Young Wynyard (2859). Young Wynyard was bred by the Marchioness of Antrim after the death of the late Sir Henry Vane Tempest, Bart., and from the old Princess, calved in 1800,

* This method of exhibiting short-horn cattle in families is now very generally adopted in America.

which Sir H. V. Tempest had bought of the late Mr. R. Colling, of Barmpton. The old Princess cow was sent to Barmpton from Wynyard, and the person who took her was told that she was to be put if possible to the bull Wellington (680), a son of Comet and out of Wildair, if Mr. R. Colling would allow her, and to pay whatever price he chose to charge. The person told me that when he took the cow to Barmpton, Mr. R. Colling told him he never bulled any gentleman's cows and could not comply with the request of the Marchioness of Londonderry. The man then left, on being refused, but had not gone far when Mr. R. Colling sent his servant after him to say if he would pay ten guineas she should be bulled; the servant saying to him at the same time, "I told my master this was not a gentleman's cow but a lady's, and that he might allow her to be bulled." Mr. R. Colling then said to him— "I believe I promised Sir Henry that if ever he was without a bull, I would allow Princess to be served by one of my bulls." So the man returned and had her put to Wellington. This I had from the man himself. I also learnt the same fact from one of Mr. R. Colling's servants, the one who went to call the man back with the cow.

I mention all this particularly to remove any doubts. The stock of Young Wynyard proved their superiority over that of Old Wynyard, though those of the latter were very good. I may also state here, that at the first sale at Wynyard, in 1813, soon after the death of the late Sir H. V. Tempest, that the Marchioness instructed the under-steward to desire one of the tenants to buy, and let her have again, old Princess, and her granddaughter Angelina, which he did, and they were kept for several years at

Wynyard, and the bull, Young Wynyard, was used to these cows, and was the sire of Waterloo and Angelina 2nd. (the sire and dam of Belvedere). He was only let out for one year to Mr. Perritt, who had been using Sir H. V. Tempest's bulls from the time he (Sir H. V. Tempest) began to breed, in 1802, and never used any other till his decease, and all the calves he had by this bull were excellent.

The Marchioness allowed other persons' cows to be sent to Young Wynyard, and I bought some cows by him and they were all extraordinary good short-horns. Some time after the second sale at Wynyard, in 1818, Young Wynyard went to Ireland, but Mr. Stephenson got his Angelina cow, granddaughter of Princess, put to him before he left Wynyard, after she had calved Angelina. I have also seen cattle bred from a cow, a daughter of old Princess, by Mr. Baker's Lawnsleeves, and I bought descendants of the said daughter some years afterwards, and three heifer calves descended from her, but I have not now any cows descended from this cow and her three daughters. I also bought, at one time and another, about thirty-seven females descended from the Wynyard blood, besides the above four I have named, bred from old Princess. This last female calf that old Princess had by her own son, Young Wynyard, was also sold at the last Wynyard sale, in 1818, when Angelina was bought by Mr. Stephenson, or, at least, she was bought of Mr. Eliker, soon after the sale, for Mr. Stephenson was not at the sale himself. Young Wynyard was calved early in the year 1815, and was used first when only eight months old, and was used up to October, 1819, and was the only bull used there. They sent Angelina in 1815 to Lawnsleeves, the only cow sent to him. In 1816 old Princess produced a heifer calf by a son of Angelina, which was by old Wellington (684). In 1817 old Princess had a heifer calf by Young Wynyard, the last calf, and at the sale of the Countess of Antrim, at Wynyard, in 1818, there were only five animals sold. Anna, by Lawnsleeves, was not sold, not being one of those five, nor Young Wynyard. Those sold were Angelina, calved in 1810, and a heifer calf off old Princess. The first was bought by Mr. John Stephenson, of White House, near to Wolverston, and was then in calf with Angelina 2nd. (calved 2nd February, 1819). She was again sent to Young Wynyard, and had a bull calf, Waterloo (calved December 26, 1819). No. 3, a white bull, by old Wellington out of Angelina, was three years old at the time of sale. No. 4, a son of Angelina, and by Young Wynyard, was calved in 1817. No. 5 was a son of Young Wynyard, out of a cow kept at Wynyard, but dam not certain. A heifer calf out of old Princess, at the sale of the late Sir H. V. Tempest, in 1813, was retained at Wynyard. She, early in 1817, produced her first and only calf, a bull, by Young Wynyard. At Wynyard she died, and this calf is believed to be No. 5. Mr. Perritt died in 1820, and had retained Anna. After that, Young Wellington or Wynyard had gone to Ireland. Young Wynyard was at Mr. Perritt's from October, 1818, to March, 1820. Pilot, from Princess, and by Wynyard, was bought in at the sale in 1813, and died the next day. Wellington, by Wynyard from Angelina, was one year old, and was bought in at the sale in 1813, and was kept till the autumn of 1814, and sold to Dawson Lambton, Esq., of Biddick.

I have given all these points thus particularly as I write from papers I have by me, which are signed by the persons from whom I had them; for, after I had bought Belvedere, I requested all the persoas who had been at Wynyard from 1810 to 1822 to be sent for, and I would pay all their expenses for time and trouble in coming to meet me at Stockton, which they all did, and these persons, so assembled (of whom Mr. Stephenson, the breeder of Belvedere, was one), all agreed in relating every circumstance that occurred from 1810 to 1822, which was after Angelina came into Mr. Stephenson's possession, and she had bred Angelina 2nd. and her brother Waterloo, the said Angelina having had only these two by Young Wynyard, but afterwards bred others, two females in succession to Baron (58), a son of Comet, and bred by Col. Trotter.

I have been thus explicit, that no doubt might exist of all the circumstances I have named; but though I came into possession of so many females of the Wynyard blood, descended from old Princess, I had only one (the Matchem cow, bought at Mr. Brown's sale in 1831) whose stock turned out remarkably well, and she had bred several calves before I put her to bulls having Duchess blood. But none of them were extraordinary; in fact, my tenants, Messrs. Bell, whose property she had become, determined to sell her, and Thomas Bell took her to Darlington market, but brought her home again. I told him I would give him the price he asked, and that if I bought her I would put her to a bull having Duchess blood, and I was very certain she would not, to that blood, breed a calf worth less than one hundred guineas. He said she led all the other cows wrong by leaping the hedges, and therefore he would sell her. I then had her put to the Duke of Cleveland (1937) and she produced a heifer calf, Nov. 3rd, 1834, which calf I showed at York in 1838 as a three year old in calf, and obtained the second premium, the first premium being awarded to another cow I showed having two crosses by Belvedere. She was a twin heifer (both breeding). When the judges went into the three years old class of cows, they immediately took these two out, but were very long in deciding which to place first, and at length asked the holders of the two cows whose they were, and on learning that they both belonged to me, and seeing me in the next pen examining the aged cows, which they had decided upon, they called me (it being then past the hour that the public were admitted into the yard) and I went to them and they asked which I considered the best? I made for answer, "I thought they need not to ask, as they spoke for themselves. The one having calved the 21st May last (a bull calf, Omega, which I sold afterwards to Mr. Foster, Springfield, Ireland), and giving a large quantity of milk, could not look as blooming as the other which had gained twenty stones in weight, while the one in milk had not gained in weight, but paid as much or more in the milk she gave, as the other had gained in fatness." They then placed the first premium for short-horns on No. 4 (the twin heifer) and the second premium on the cow off the Matchem cow, by Duke of Cleveland, and did so properly under the circumstances. It is rare, indeed, that judges at exhibitions of stock give premiums to great milkers: they look only at the condition or fatness of the animal. This cow, off the Matchem cow, produced a bull calf, named Locomotive (4242) Oct. 5th, 1838, which I sold the March following to J. C. Etches, Esq., then of Liverpool and Barton Park, near Derby, at 100 guineas, and after using him some time he sold him to Mr. J. C. Letton, Pincey Grove, Citron Forest, Kentucky, U.S., America, at 250 guineas. Next year, 1839, I showed this cow (afterwards called the Oxford Premium Cow) at the meeting at Oxford of the Royal Agricultural Society of England, and she obtained the highest premium as the best cow (though only second to one of my own cows at York, as a three years old). The same year, on the 24th Oct., 1839, this Oxford Premium Cow produced another bull calf by Short Tail (2621), named Duke of Wellington (3654) which was afterwads sold to Mr. Vail, near Troy, State of New York, America. These two bull calves obtained the highest premiums in America, and their sons also have been equally successful, and Locomotive's son Walton, belonging to Mr. C. W. Harvey, near Liverpool, obtained the highest premium at the Beverley meeting of the Yorkshire Society in 1845. Afterwards in the same year, he and his son, Earl of Sefton, obtained the highest premiums at Liverpool, and Walton was shipped the same evening to be shown at Dumfries, and was again successful in obtaining the highest premium given by the Highland Society of Scotland held in that town.

This same cow also bred a steer, own brother to the Oxford Premium Cow, that at two years old weighed 72 stones, a weight that surprised me, as I sold him to weigh 64 stones, but the butcher who killed him said he expected the steer would exceed my estimate, as he had bought four steers a year older of mine in 1836, and they far exceeded his own estimated weight of 90 stones each. I asked him to what cause he assigned their weights beyond his expectations, and he said it was the great width of their backs, and their great fatness all over, being uniformly good in every point. The same cow has bred two bulls, Cleveland Lad, and 2nd. Cleveland Lad, which are both living, and very active, though one is now ten years old and an own sister also, which has produced three bulls and two heifers, all excellent. Cleveland Lad obtained in 1841, the highest premium, as the best bull at the Royal Society's meeting held at Liverpool, and at the Yorkshire Society's meeting at Hull the same year. 2nd. Cleveland Lad obtained the second premium at Doncaster in 1843, and the highest premium in 1844, as the best bull at the Yorkshire Society's meeting at Richmond.

I name all the above facts, not to blazon my own breed of short-horns before the world, but as a duty I could not avoid, as I am confident it is to the animals I have named that future breeders will become indebted for having the best breed of short-horns, as I am convinced they will go on improving every generation, whatever stock they may become incorporated with, whether they be short-horns or the other valuable breeds abounding in the United Kingdom.

From 1790 to 1807 I weighed food both to milch cows and to feeding cattle to ascertain the food they consumed, and measured the animals to ascertain the improvement made, as I found I ascertained the improvement made better by measurement than by weighing them alive at different periods. But, as the tribes of stock I more particularly ascertained the merit of by the above test are not in existence, excepting my Duchess family, I shall not enter into particulars. I cannot, however, avoid recommending all persons to examine their stock by this only true criterion, for it was thus that I came to know the great difference between different varieties of stock, and learnt by the external characters of each their real merits, for the same data extended to others proved a like result.

The late Mr. Thomas Ratcliffe, for more than twenty years the principal butcher at North Shields, then lived at Newcastle, and commenced business in 1790, and for some years he came to me and bought two beasts, taking one that week and the other the week following. He was a well educated, intelligent young man, and very attentive to business, and we compared notes. I told him the quantity of food consumed by each, and the improvement in weight, and he informed me how the animals died, their weights, &c., which mutual communication improved the judgment of each, and he often informed me of the benefit it was to him in improving his judgment. I remember on one occasion seeing six cattle killed one morning, and I had put down what I considered each would weigh, and on being weighed the following day I was very few pounds wrong in the weight of any one, and hardly wrong in the whole six cattle. But I exercised my own judgment in addition to the measurement, as the difference in the form, degree of fatness, &c., causes a difference in the weight which the measurement can not give accurately. The judgment must, therefore, be exercised in addition. It was the long perseverance in this plan of weighing food consumed, and in ascertaining the improvement of each by measurement, that matured my judgment, and enabled me afterwards to decide accurately on the merits of animals by handling and external appearance. This practice, when once learnt, is never forgotton, but remains as fixed in the memory as the pence or multiplication tables learnt at an earlier period. It is necessary to state here that the breed of short-horns in Tyneside were, from 1785 to 1795, when I first examined the stock accurately, very inferior to what they afterwards became; many were like the Ayrshire cattle which prevailed in that country when I examined them in 1805.

PART II.

THE QUALITIES OF MONGREL AND PURE-BRED SHORT-HORNS DISCUSSED, AND THE VALUE OF EACH COMPARED.

Coarse-beefed Short-horns Described—Mr. Culley's Evidence as to the Introduction of Coarse Dutch Animals—Hubback's Influence on the Issue of Coarse Cows—Durham was not free from the Importation of these Inferior Stock—Mistake by Dairymen in regard to the Richness or Quality of Milk—Consumption of Food by Different Animals Tested by Mr. Bates—St. John's Delicate Constitution further Proved—The Loss of Ketton Appreciated—The Earl made a Great Mark—The Three Tribes to which the Bates Short-horns Owe their Excellence—Mr. C. Colling and the Introduction of the Galloway Strain to his Herd of Short-horns; the Effect being "Hard Flesh, Bad Wiry Hair, and No Milkers"—The "Fool's Piece," again—One Cross may be Superseded, but Two Crosses, never—The Value of this Knowledge—Judges at Shows Incompetent, or Worse.

The coarse-beefed short-horns so accurately described by Mr. George Culley, in his treatise before referred to, were very common when he wrote. What he states of them, too, is very true. "They were black fleshed; for notwithstanding one of these creatures will feed to a vast weight, and though fed ever so long, yet will not have one pound of fat about it, neither within nor without, and the flesh (for it does not deserve to be called beef), is as black and coarse grained as horse flesh, and," as he justly observes, "no man will buy one of this kind if he knows anything of the matter, and, if he should be once take in, he will remember it for the future." He says, further, "I once saw a beast of this sort killed, which, after feeding all summer, had not a pound of flesh inside or out, and was more like an ill-made black horse than an ox or a cow."

In the spring of 1794 I bought some of this kind of cattle, having been tempted by the low price, and the same day I also bought some other steers (each above three years old), but very superior good grazers. They were kept well on turnips for three months, but the latter increased double what the former did, while they consumed far less food; so that the latter left more than four times as much for the food consumed as the former, which cured me for ever of buying another of the same kind.

Mr. Culley states that these coarse-beefed cattle were introduced from Holland by persons going over there after Dobinson had introduced the good variety from Holland, but who lacked Dobinson's judgment. I have not seen for many years any of these very inferior short-horns, though I know some short-horns that were originally of this variety. But by crossing with a son of Hubback they became much improved. This has been going on for above fifty years, and they have since bred, and for some years have been incorporated into herds of high note, but in which I can perceive a degree of resemblance to their originals in some of the descendants, and on which account no consideration would induce me to have one of them to breed from. There were also importations of a very large variety of short-horns, and I have been told they were introduced from the Continent by the Haggerston family, in Northumberland, and prevailed above forty years ago along the coast of that county. These I have frequently bought and grazed; they were free from the coarse-fleshed variety, but were immense consumers of food, when fed on turnips after a summer's grass. They never paid for the food they consumed, although they were not bad growers like the coarse-fleshed variety; but, from the crosses of Messrs. Colling's cattle extending into Northumberland, even this variety were not what they once were. This same breed may have been introduced into the county of Durham, when size was so much run upon, before Mr. C. Colling exhibited the bull Hubback. The character of Mr. Hill's cattle, of Blackwell, resembled in degree these large short-horns that I have mentioned as being in Northumberland; and when I visited Lincolnshire, in 1810, I found that large variety of short-horns very prevalent in that district, particularly on the good land about Boston, where there is a quality of grass land deep and rich, and not unlike that around Haggerston, in Northumberland. Those who bred cattle for the dairy principally, and did not rear steers to feed for the butcher (which much prevailed in my early years, both in Northumberland, Durham, and Yorkshire), selected such cattle as gave the greatest quantity of milk, without paying regard to the richness or quality and quantity of cream and butter. These were very slow feeding cattle when they became dry, and they were sold at low prices. But such like dairy cattle are not now kept to the extent they formerly were, though many who attend to the dairy alone still have an inferior grazing stock when made dry to feed off for the butcher.

Finding so very few really good grazing short-horn cattle to breed from when I took Halton in 1800, and having found that the best West Highland cattle were small consumers, and good grazers, I bought the best I could obtain to put to the bulls I hired of the Messrs. Colling. Their offspring I reared, and continued to breed from them for twenty-five years. In 1800, I obtained twenty, and in 1801, twenty more, which were by far the best I ever knew of that breed. They were selected by a good judge, according to my directions, from many thousands purchased in the Highlands during these two years; but after them I never could get any more comparable with them, in fact, ten out of the forty were nearly equal to the best short-horns, and by the cross with the best short-horn bulls, their issue became superior to any short-horns except the Duchess blood. They gave also very rich milk, and the cross equalled the short-horns in quality of milk, and retained the rich quality of the West Highland cattle.

As regards the feeding of these cattle, while my first Duchess consumed thirty-six pounds per day of hay (the smallest consumption of any short-horned cow I had then ever known), those bred between the short-horn and the West Highlander, consumed only twenty-four pounds per day, which was very little more than was eaten by the West Highlanders themselves. The steers so bred, and fed on a like consumption of food, were nearly one-third more weight at three years old. They were equally as great growers as the best short-horns, and worth more per pound in the Skipton market at that time than the short-horns, and equally as much as the best West Highlanders. Each cross I made kept improving them, and I should never have relinquished this breed had I not had the misfortune to introduce St. John blood. No cattle were ever more hardy than they were previous to that cross, but those so bred inherited the delicacy of all the blood of St. John, and though I kept them for above twelve years after I had introduced the St. John blood, yet I could never recover their constitutions again, and was obliged at last to part with the whole of them. Since then, 1826, I have bred nothing but pure short-horns. All the short-horns into which I introduced the St. John blood partook of the delicacy of that tribe, though nothing could be more hardy than they were before such introduction, and all these short-horns so bred I was obliged to part with at an immense loss, and such has been the case with every herd where that blood has been imported.

I should mention here that, previous to Mr. C. Colling's Hubback, the prevailing characteristics of the best short-horns were high hooks, and the fore-quarters small, and very lightly lyred, as short-horn cattle so shaped were the very reverse to the objectionable coarse-fleshed variety. But Hubback gave a fore-quarter much heavier, and the hooks, though not so high, were wide and pointed forward, making a much longer hind-quarter. This was particularly the character of my cattle, bred between the best short-horns and the West Highlander. When I began to use my bull Ketton, the produce from him made both my short-horns and mixed blood very superior to what they were from the Daisy bull and Styford, the best two bulls I had used before I had Ketton. When Ketton ceased to procreate, I then felt severely the contrast, even with the best which had descended from him. I then did not know that any of the Princess blood was left, nor did I know how the Princess tribe was bred, direct from Hubback to Favourite. This I did not learn till I made enquiries to assist old Mr. Coates in preparing the first Herd Book, and I saw no way of restoring the pure Duchess blood but by putting the 3rd Duchess to her mother's own brother, Duke (226), sold at the Ketton sale in 1810. This I accomplished in 1820. The Earl (646), so bred, was calved in Nov., 1820, and putting him to the Acklam Red Rose cow produced my second Hubback (1423), January 25th, 1823. This union answered most admirably, and gave me a uniformity of character possessing every excellence. I used second Hubback till I got Belvedere in June, 1831. My best stock are descended from second Hubback's daughters, put to Belvedere, a union that answered again most admirably, so that I have combined in my breed of short-horns the only three tribes that were all bred direct from Hubback to Favourite, and there are no short-horns that have now any pure blood of any of those three tribes, excepting my own herd.

It is the union of those three bloods, Duchess tribe, Yarborough's dam, and Princess tribe, to which my cattle owe their superior excellence. Having united these three tribes through so large a herd as I and my tenants, the Messrs. Bell, possess, and those who have got the blood through my herd, extending through six tribes in my own hands, and eight of my tenants, we have every means, not only of preserving their excellent qualities, but of improving them more and more every fresh cross. We are not dependent on any one bull, as every cow has more or less Duchess blood, with all the excellencies which cattle so bred possess. In fact, as Duchess 34th proved, by several trials repeatedly made in various years, she consumed less food (hay), per day than my first Duchess, grandam of No. 1, in Herd Book. From a quart of milk, the yield of butter from Duchess 34th was more than from my first Duchess. These are important considerations, when weighed in conjunction with a greater increase of weight in beef, with so much less consumption of food. In fact this consumption of food and production of meat were nearly equal to the small consumption of the West Highlanders, which are the smallest consumers of any cattle I ever tried on hay alone in winter; and I have ever found the consumption of hay correspond with the consumption of turnips or straw given singly.

It is well known that Mr. C. Colling unfortunately introduced the Galloway or Polled breed of cattle into his short-horns, through a cow of that breed having been sent to his bull Bolingbroke. He stipulated that if it should be a bull calf he was to have it. This bull calf, so bred, was again put to a short-horn cow of the Johanna tribe, which were very hard handlers, and slow feeders. They were of a neat form, but gave little milk, the cows of that tribe all partaking of these bad qualities:—hard flesh, bad wiry hair, and no milkers. Those kept on forcing food, looked well to the eye of men who did not estimate properly, for they caught the eye by showing the fat on the tail, which, as has been said, is called the "fool's piece." Of this second cross through the Johanna cow, I saw a bull shown at Durham, 31st March, 1797, when he was placed second. He was the same age as the Favourite bull, and Mr. C. Colling was so imprudent as to put this bull to Favourite's dam, and the produce, Lady, introduced this blood into many herds of short-horns. Lady had of course one-eighth of this blood, and her next descendants one-sixteenth, and so kept decreasing each cross. Where only one cross has taken place, the previous character may be restored by putting them to the very best short-horn bulls; but where two crosses have been introduced, *I never have seen any animal* so bred that could produce a good bull. It is valuable to know that this is the case, as it checks their spread where this is seen. But, unfortunately, many herds have got three or four or more crosses of these mixed mongrels, and such cannot now be restored. Their owners, however, still persevere in producing these mongrels or nondescripts. And from men having such being judges at cattle shows, the premiuns are given to such animals to help to holster up the character of their own herd to the public detriment. This has ever been the case since this mongrel blood has become so prevalent— particularly since Mr. C. Colling's sale in 1810—to the present time.

PART III.

MR. BATES COMMENDS HIS PURE HERD TO POSTERITY.

Gambier and Belvedere purchased—The Value of Pure Short-horns of 1845—The principle which guided Mr. Bates; he studied the Fame of his Herd more than his Fortune in Life—Mr. Bates on Crossing Breeds—The Highland Herds and Pure Short-horns—Highlanders made more Hardy by the Introduction of Short-horn Blood—Stock that will not bear the Test of Ordinary Hardships must be Unprofitable—Mr. Bates Weeded his Herd of Faulty Cows, as for Abortion and other Defects—Bolstering up Degenerating Families—What "a Painter without Principle" may do—The Value of Pedigree, and in what this Value consists—Mr. Bates's Object in Writing this History—Hubback, Strawberry, and Strawberry's Sister—The Herd Book begun and continued with False Entries—Mr. Coates's Confession, and his Poverty his Excuse—Lily and Fortune Objectionable Cows, and the St. John Bull and his Weak Descendants—Jupiter Coarse and Ill-bred—Mr. Mason admitted losing £20,000 by his Herd, which is Fully and Intelligibly Accounted for by Mr. Bates—Mr. Bates Insulted for being Candid and Truthful; his Assailant Apologizes; why this is referred to—The Galloway Scotch Blood again Noticed in Connection with Bolingbroke—Mr. Bates bought the First Cow Mr. C. Colling sold for 100 Guineas—Comet Valued at 1500 Guineas when 7 Years of Age; Died 1815—Ketton Improved the Comet Issue.

I have thought it my duty to say this much, and, having fairly tested the merits of the best of those so bred, I shall give the result to guard others against so fatal an error. In 1831, I purchased Gambier (2046), so bred, of Mr. Whitaker, and soon afterwards I purchased Belvedere of Mr. Stephenson. I had eight steers so bred to each bull. Gambier was by Bertram, which had the Daisy blood, one of Mr. C. Colling's best tribes of short-horns, and was from Lady Matilda, the highest priced cow bought at Mr. Charge's sale, in 1828. I paid double the price for

Gambier that I paid for Belvedere, having both bulls in use at the same time.

Mr. Bates then detailed the results of the feeding of the two lots, and the superiority of those by Belvedere. In the above contrast there was a test of value which there was no evading; and the result thereof must carry conviction to everyone whose prejudice is not predominant, but who is in search of truth.

I cannot conclude these details of the short-horns without apologising for such a desultory production; but the facts I had to state, and the nature of the subject, have caused it to be so. It is a subject on which my attention has been engaged for above sixty-five years, and is therefore the result of a long experience. I trust it may help to awaken the attention of landowners as well as tenants, as the value of landed property depends so greatly on the value of the cattle depastured or wintered thereon. Whoever begins breeding short-horns after the little information I have here given, may proceed on sure grounds.

Every one by trying the cattle they are possessed of may, by the same test I have made, see what advantage they may gain by possessing the same blood, or, by assimulating it with their present stock, and seeing the result, proceed accordingly. The several varieties, the result of the original cows being different, will enable each one to select, according to his judgment, whichever he may approve or wish to make a trial of. Then, again, a change to another variety may make a still further improvement in a herd. There is in my view no bounds set to improvement, if breeding be only judiciously carried out, which of course depends upon the judgment exercised. To point out any particular line on which to proceed, would depend on what stock persons possess when they begin to cross with the best blood.

I have hitherto retained my herd entire to prove their superior excellence. One grand result I hope will follow; the herd will be able to produce stock for supplying new beginners with proper animals for a commencement. I have also given a short detail of the stock I bred between West Highland heifers and short-horn bulls, hoping that breeders of West Highland cattle may be induced to follow what I have found to be so beneficial, and I trust it may induce breeders of other cattle, Herefords, Devons, Sussex, and what are called home-breds in Norfolk, as well as short-horn breeders, to try the result of a cross of this blood. The breeders of Aberdeenshire cattle have felt the benefit of a cross with what too often were very ordinary short-horns.*

If they try it with the best breed of short-horns, they will find the result very superior to their past trials. And if they begin with the originals of the Aberdeenshire cattle, the better the short-horn bulls are so much greater

---

* Mr. Bates always repudiated the common idea that any thing was good enough to cross with. I am informed by an experienced Scotch agriculturist that many breeders have tried, with success, the well known fact in horse breeding, viz.:—that a mare will generally produce what takes after the sire of her first foal, although the subsequent sires may be of a totally different breed or class. Thus they put the heifers to a first-class short-horn bull, and then breed from the pure bulls of their own race, and the produce will retain the excellence of the short-horn, though the characteristics and colour of their own breed is retained. In my experience I have found that if heifers are put to an inferior bull, it is a long time, before, if ever, that they will afterwards breed good stock when put to good bulls. —*The Author.*

will be the benefit of the cross. This is more absolutely necessary in crossing breeds than even when breeding from pure blood, for if there is any coarseness on either side, the stock is sure to inherit that coarseness; but when free from coarseness, there is then no fear of crossing different varieties, but *always let the male animal be the one to improve on the female*, that is, let the male be of a superior tribe.

I can have no doubt whatever that the best short-horns will improve all other breeds to which the *male animals* of the short-horns are put, and when once crossed, the owners may then increase the short-horn blood if they think proper, or breed between males and females of this mixed breed, or use those so bred to the best of their original breed, whether Hereford or Devon, or any other breed of cattle. But I am convinced that with the West Highland heifer and the short-horn bull the greatest improvement may be made. From all the trials I made (till the unfortunate cross with the St. John blood), the increase of short-horn blood did not diminish the hardiness of the highlander, but the contrary in every instance. This was more particularly the case with the cross by my Ketton, on the daughters and granddaughters of the Daisy bull, the best cross I made till I used Ketton. I had an open pasture of 200 acres on what had been Broadpool Common, part of the Park End property, very high lying land, in the west part of Northumberland, near to Cumberland. It was near the summit level, where the waters fall into the east sea near Tynemouth, and the west sea on the Solway Firth. There are few places lie higher in Scotland—except the mountain scenery—between their east and west seas. The winters are longer in the high parts of Northumberland than in Scotland, and the weather is more severe; yet I found the stock bred between the West Highlander and the short-horn equally or more hardy than the pure bred West Highlanders. I tried them in that exposed situation from 1802 to 1807, inclusive, six years, and found the same result every year, both in winter and summer. The cattle were tied up at nights in byers in the winter, and went out in the day to top the heather and eat the coarse herbage that grew on this exposed situation. The half-bred, or mixed blood, I have named, had a double coat of hair, one long, and the other a close fur or down, thicker than that of the highland breed. I sent the calves to this situation before winter; wintered and summered them there two years; and at two and-a-half years old brought them to the tillage farms I had lower down in the country. I there gave them straw and a few turnips the first winter, then summer grazed them on inferior land till the autumn, when they got fog or aftermath; the hay having been cut and intended for the milk cows after calving, in April and May. No other cattle ever had hay with me in Northumberland, unless it was a little coarse hay from bog land. And the last winter, when three and a-half years old, they got turnips, and were sent to market in April, when I obtained from £26 to £27 per head for them in the years 1805 and 1806; beef was then about 7d. per pound for the very best highlanders, and the ordinary short-horns not more than 6d. per pound, or even less.

The land I name on Broadpool was rented in 1794 at 1s. per acre, and after partially surface drained was valued at 3s. per acre when sold at the highest time of the late war, in 1806. The cattle did not cost keeping from half-a-year old to two and a-half—two years—more than £3 per head, as I kept 24 head of cattle there winter and summer, besides the shepherd's cows, which he had the liberty to keep for mowing the hay on 20 acres and winning it, and attending to the cattle. When this small cost in rearing cattle for two years is deducted from the price they were sold for, and their dams amply paying in milk and butter for the land they went on, besides keeping their calves till half-a-year old, the return for the next year (between 2½ and 3½) was ample at £3 to £4, leaving £20 for the last winter's keep on turnips and straw, which shews what such cattle may do. But in after years, when I ceased to have this high lying land, and sold the cattle in May and June off grass when three years old, at £37 per head, it shews how much greater was the return. But beef then rose to 12s. per stone in 1813, and the steers averaged 62 stones each, when under three years old, and their keep till the last year did not cost me one-third of the price they were sold for, paying near £26 for the last year's keep and above £20 for the winter's keep. I have given this long detail, that breeders may see it was not by forced keep the cattle, so bred, were made to be sold at high prices, but that it was from the excellence of the breed. It is necessary my readers should ever bear in mind, that I have never at any time of my life kept my breeding cows and young stock high, and only gave them when finishing for the butcher such keep as they amply paid for. This they did more than any cattle I could buy in to feed for the butcher, and they would leave more profit or a greater return for the food consumed than any other cattle, not my own breeding, would pay.

Stock that will not bear these tests are ever unprofitable to the breeder, and this has injured, and too often ruined breeders of short-horns, as dear bought experience has proved to too many of them. Some of these men, however, railed at me for not keeping my cattle better, saying "persons could not judge of my cattle as I kept them so poorly when young, while they, by giving their calves wet-nurses, and afterwards high keep, shewed them much forward." Now I have seen the end of many such, who came at last to have no short-horns left worth looking at, while mine have remained unimpared in constitution, and kept on breeding. As a proof, I have now my 58th Duchess since Mr. C. Colling's sale, besides many not numbered, and scarcely ever keeping a cow over year, when not in calf in the autumn. As a proof of this, I parted with the grandam of Duke of Northumberland when three years old, because she slipped her second calf. I made her dry and sold her fat in two months afterwards. Several others of the same age, of excellent quality, I similarly disposed off, lest that complaint should again occur in my herd, having had in 1804-5 a severe loss, by cows casting their calf, when I used Mr. R. Colling's bull Styford in 1804. Had I kept on such cows, my Duchess would far have exceeded the number of 60 in the period from Mr. C. Colling's sale in 1810; and yet, after all, where is there one breeder of short-horns who can show one-tenth of the number since that period, from the animals that were bought at that sale? I doubt it much whether there are *any* now remaining, and if so, who has them, or where are they to be found? Not in Coates' Herd Book at least. And would they not have been there had there been any? As to the Chilton herd, except it had been for the cross with St. Albans,* where would any have been found? Nowhere, except in the coarse ill-bred strain of Jupiter, and his sons, Mars, &c., &c., whose progeny shewed a character, as breeders, that it took even St. Albans to restore. This was after Jupiter had been used on the tender delicate Sir Oliver's progeny, and their relatives from the St. John blood, where the delicacy commenced, and if it was not forced keep that brought that delicacy on—what was it? That it became hereditary there can be no doubt now; but never before St. John's day did I ever hear of hereditary disease in cattle. I well remember in August, 1812, when Baroness, by St. John, the daughter of my first Duchess, dropped down in the byer after milking, and never rose again. I sent for a cow doctor (Mr. John Charlton, of Heddon-on-the-Wall), and, on seeing her said, "this is what I have met with before in cattle having the St. John's blood, and I am certain that it is hereditary in that blood." I had never heard of hereditary disease before in cattle, and I laughed at the idea, but dear bought experience convinced me of the reality, and many who had that blood suffered the loss of thousands of pounds therefrom. This

---

* This was Mr. John Wood's bull (1412). The story is well known, and I believe authentic, that Mr. Wood sold his bull to a butcher to slaughter, and that some years after in looking over Mr. Mason's farm he found a bull in a dark stall, and on looking at him at once recognized his old bull, which he supposed had been killed long before, and to this bull the excellence of Mr. Mason's stock afterwards was due. Mr. Mason bought him from the butcher, or had sent the butcher to buy him from Mr. Wood.

they have acknowledged to me, while there is the confession made by St. John's owner, Mr. Mason, at his sale in 1829, that he had lost more than twenty thousand pounds by breeding short-horns, that could not by any stratagem be evaded. Yet, strange to say, in this day, after Sir Thomas Fairfax blood can no longer be bolstered up as unsurpassable, the very dregs of Mr. Mason's herd are brought forward and held up as non-suches; or, to use their own words, "proved by authority" the "invincible." Was ever one of that tribe heard of till Cassandra and her produce were puffed up? Even Captain Shaftoe himself was placed nowhere at the Highland Society's Show, at Dumfries, in 1845, for a grandson of Duke of Northumberland, Walton, obtained the highest premium, and Mr. Jobson's bull, by second Duke of Northumberland, was placed second. A painter, without principle, may make the worst of animals to be all perfection to appearance on canvas, when the animal itself is the very reverse of anything good whatever, and only the result of forced feeding, which to the eye of a real judge, shews the deficiency to be, altogether, the more glaring.

Now the value of pedigree depends not upon the length of such pedigree, *but on the length of time there had been a succession of the best blood, without any inferior blood intervening, in such succession.* This can only be known by those who have known all the various crosses there are in any animal, and the blood or breeding of such from their own knowledge of the tribes from which they came, and how that produce bred again. There even the greatest judgment is required, which nothing but a long experience can give; and, in the present day, there are few indeed who can give that proper information which is requisite to decide with accuracy in all cases.

I have ever recorded the opinions of older breeders than myself from 1782, on which I have exercised my own judgment, and marked the results, which extended experience gave me, and as these have all agreed, after 65 years' experience, there is then no danger of error.

I have made these remarks on the stocks of the Messrs. R. and C. Colling, and shall now record my observations on others. The late Mr. John Maynard, from whom Mr. C. Colling obtained the predecessors of Comet, gave me under his hand, dated the 13th of March, 1820, a statement of many facts little known amongst breeders of short-horns in this day. He remembered that tribe of which he sold to Mr. C. Colling his cow Favourite and her daughter, Strawberry, from the year 1750, now a century ago, and he said the originals were great milkers. The first three in succession had always to be milked before calving. The distemper prevailed amongst cattle called the murrain,* when he was a boy going to school, before the year 1750, and he remembered some of his father's cattle being sold which died of that distemper; and also that this complaint did not extend beyond the river Tees to the northward, as there was a strong cordon drawn, and a strict guard kept, that no cattle should pass the Tees. He remembered besides that their servants conveyed the cream across the Tees in a boat, and churned it on the Durham side of the river, and then sent the butter to

---

* This was in fact the Rinderpest Cattle Plague of 1745, which destroyed so many cattle in the North of England and all those in in Cumberland.—*The Author.*

Darlington market, as no butter was allowed to pass the Tees for that market. Mr. Maynard's father lived at Eryholme adjoining the river Tees, and was thus enabled to dispose of his butter.

These three generations of short-horns were, *before the cross by Mr. Jolly's bull*, bred, as I have before stated, by Mr. Wm. Wastell, of Burdon; and his cow—daughter to Jolly's bull—was called Strawberry, which gave a very superior character to this tribe of short-horns. He then put the Strawberry cow, so bred, to Mr. Jacob Smith's bull of Givendale, near Boroughbridge; this bull was of a yellowish red colour, with a white back, white face, and white legs to the knee—but he knew nothing of his breeding—and this again was put to Mr. Ralph Alcock's bull (bred by Mr. Michael Jackson, Hutton Bonville, near Northallerton, the pedigree of which cannot be traced). He was remarkable for his handling, lively looks, and all his stock were like himself. This was the cow Mr. Maynard sold to Mr. C. Colling, the 30th September, 1786. She was then seven years old, and the price was £30, and twelve shillings was returned: her calf he was to take the next spring, February 13th, 1787, and her price was to be ten guineas. Mr. Maynard had an own sister to this calf, but it took still more to its sire, called Dalton Duke (kept by Mr. Maynard and Mr. Wetherall), which bull Mr. Maynard (as well as the Messrs. Colling), described as a bad haired bull, and bad handler. The year following (1787), Mr. C. Colling allowed Mr. Maynard to send three of his best cows to Hubback bull, and he had a bull calf—which partook of Hubback's character—off one of them, and on putting this bull, by Hubback, to Strawberry's sister, the produce was quite the reverse to the dam, "taking Hubback's character, and she was of a large frame, and was a great feeder, and her handling good." I give this in Mr. Maynard's own words, that it may be seen how great a change Hubback produced on this tribe of cattle; for the sister to Strawberry "had a thick skin, and was a hard handler, her hair as strong as pig bristles, and was plain shaped, and low sided," and being tired of keeping her, *he sold her to Mr. C. Mason in* 1796, *having put her dry to feed for the butcher.*

Messrs. Colling were both there at this time, on their way from Northallerton fair, and they could not think (so bad was she), that she could be a daughter of Mr. C. Colling's Favourite cow; but Mr. Maynard assured me it was the case. Mr. C. Mason on buying her, put her to Mr. C. Colling's bull, Favourite, and gave her the name of Miss Lax. Mr. C. Mason's Lily was the produce, yet she was far from being a good handler, and her hair was thin and short, though she was always kept in the very highest condition. Mr. C. Mason's other cow, kept together with Lily, in the same high condition, was named Fortune, and in the Herd Book has the following pedigree, "bred by Mr. C. Colling, calved in 1793, by Bolingbroke, dam by Foljamb, g. d. by Hubback, gr. g. d. bred by Mr. Maynard."

Now, I think it right here to state that both Mr. Maynard and Mr. Charles Colling assured me that they never had any bargain for a female short-horn, except the cow and calf above stated. Knowing this fact from both the above gentlemen, and having seen this pedigree exhibited, I mention it in order to prove the absolute necessity for all pedigrees intended for the Herd Book, to have been proved before a Committee, for that purpose, before it was published.

I named this to Mr. Jonas Whitaker, and three other persons at Mrs. Wright's sale at Cleasby, in 1820. I then desired them to accompany me to Mr. C. Colling, to ask him the question. They did accompany me, and on naming it to Mr. C. Colling, and asking him what cattle he ever bred of that pedigree, he said there could be none so bred. He added that he sold Hubback and Foljamb the same autumn, in 1787, before Foljamb was a year old, and that he only had Favourite cow, bought of Mr. Maynard, and by Hubback, in 1787, and on calving had her and her daughter served by Foljamb, before he left after being sold; that no cow of Mr. Maynard's could have both those crosses, Hubback and Foljamb. I told the party that, if a Herd Book published all that persons sent to the editor, it was of no value as a record, and only would tend to deceive. Yet the Herd Book was so published, and is, in consequence, of no value, and only tends to deceive the ignorant, and mislead them, and it can only end in injuring a cause it is presumed to benefit.

I had spent much time, and put myself to much trouble and expense in order to furnish Mr. Coates with materials, without which he could not have succeeded in publishing the first Herd Book. I think it also right to state that I procured from Mr. Alexander Hall the full pedigrees of the Princess, or Wynyard tribe of cattle, while he had them in his possession, and also from John Chapman (Mr. R. Colling's man; Mr. R. Colling having died that spring, and left no written documents of the pedigrees of his short-horns), as a letter from Mr. C. Colling to me shows, which statements of John Chapman I had repeatedly heard from Mr. R. Colling, and particularly on the last occasion when I saw him in the autumn of 1819, on returning from Mr. Witham's sale at Lartington.

In fact, Mr. Coates was influenced by parties who got him to do whatever they pleased, to puff off their stock; but which statements ought never to have been entered in the Herd Book. And, but for my accompanying Mr. Coates to Mr. C. Colling's, with the written documents I had preserved, given me by Mr. Colling some years before, and knowing where he kept his memorandums, that they might be referred to, the old pedigrees could not have been obtained; even some of these were altered by Mr. Coates. On reminding him, when the Herd Book was published, that Punch's pedigree and his sire's, which are put down as having two crosses by the same bull—whereas Mr. C. Colling positively told Mr. Coates, in my presence, that neither he nor his brother ever put a daughter to her sire till they used Favourite bull—Mr. Coates admitted it, but said, "I had some good friends that had this blood, and they wished to have good pedigrees to them."

I think it also my duty to state, that not only were Lily and Fortune cows objectionable in their original descent, but being kept too high in condition, or from other causes, their constitutions were injured, and the greatest losses were sustained by those who used that objectionable blood through St. John bull and his descendants, Sir Oliver, Pope, Ladrone, &c., &c., &c. Nor can any safely be had that it will not show itself in the progeny however remote in degree.

That coarse ill-bred bull, Mr. Mason's Jupiter and his descendants, were used to try to do away with that delicacy, as was other coarse ill-bred tribes introduced by others who bought that blood at Mr. Mason's sale for the same purpose. But as a proof of the superiority of the Hubback blood when introduced to these cattle which had lost character, I may mention the St. Alban's blood, as seen at Mr. Mason's sale in 1829: for those with this cross showed it in their hair, handling, and looks, very different from what Mr. Mason's cattle were before that cross, And I have been told that animals so descended lived and bred very different issue to what was done before that cross of the Wynyard blood, although he affected ever to hold that blood, till then, in no estimation. But his sale showed the contrary, both in the appearance of the cattle and the prices which they brought. Who, however, that knew the loss that had been sustained previously, would run the risk of using such stock again? Only strangers thereto, would do so, unless they wanted to bolster up the animals in their possession that they may make the best of what they hold, however unworthy they have found them from dear-bought experience.

I need only remind those who dined at Chilton, after the first day's sale, in 1829, of what I had often heard repeated. Mr. Mason "drank the health of his party, and wished them success of their bargains, but he must admit to them that he had lost twenty thousand pounds by breeding short-horns." Of this fact I could have no doubt whatever, as I had often told him so many years before that time, for keeping one lot of cows to breed, another to suckle the calves from the first lot, and a third to give milk for the house, was what few could afford to do, be their circumstances what they may.

I cannot complete a true history of the short-horns, without stating another fact regarding a tribe of cattle now intermixed with Mason's old blood. The predecessors of Firby (1040) are stated to have the blood of Styford. Now, I not only know it as a fact, but told the owner at a Doncaster meeting (in 1820, I believe) of his error in putting Styford in that pedigree, and saying at the same time, that Colonel, which is placed as the next cross, was slaughtered after being used some years at Styford (Mr. Thomas Jobling's farm in Tyneside), before Styford bull came there in 1799-1800. Mr. Jonas Whitaker and the Rev. A. Rhodes may well remember this circumstance, from what they saw take place at the said Doncaster meeting, after I had told the owner of his mistake. He certainly apologized next day, for his conduct in the presence of the two gentlemen I have named, and they cannot well have forgotten the circumstances. I state this as a duty to the public, to let it be known that, in writing this history, I am well aware of the offence that will be taken at this and many other statements I have had to make. But truth required it from me, and I have on that account given it.

I think it the more necessary to do so at a time when a large party are attempting to bolster up Mr. Mason's blood, as if they were super-eminent, when they have failed in bolstering up the improper Galloway cross—introduced by Mr. C. Colling to his short-horns, when he used the grandson of Bolingbroke (280), as he is called in the 1st vol. of the Herd Book. This is the tribe that not only introduced the Galloway Scotch blood, but the hard fleshed Johanna tribe, which Mr. C. Colling hated, and he never concealed from me his dislike to that hard flesh. But their shape ran away with breeders who knew not the *estimable requisite*—good handling, without which there can be no good short-horns, however bred. Many suppose, because others have asserted it, that Mr. C. Colling introduced this cross to improve the short-horns, when the very reverse is the case. It was accidental at first, Mr. G. Coates having bought two Galloway hornless Red Scotch cattle for Colonel O'Callaghan.

In March, 1800, I purchased my first short-horn cows of Mr. C. Colling, and agreed to give him the first 100 guineas he ever sold a cow for. He sold many afterwards, however, for more money. One, own sister of the one I bought in 1800, he sold in 1806 at 300 guineas, to General Simpson, in Fifeshire. He let, when a calf, at the same time, North Star, for two years, and received 200 guineas for his use, although the General then had only a very few cows to put to him, and he was going into a part of the country where no rivalry could arise. The General would never allow the bull to return, remitting 100 guineas annually for his use. In this I think he was wrong, as Mr. R. Colling had engaged the bull of his brother before Mr. C. Colling's sale in 1810, before Comet was sold, and, of course, relied on having him home that autumn, whereas the General would not part with him, and he died the following spring, at the General's, at 5 years old.

After the death of North Star, the owners of Comet where offered 1500 guineas for him, though then turned 7 years old. Comet died in June, 1815. Mr. James Fawcett, of Scalby Castle, took a faithful likeness of Comet, which I have here, the week before he died, on the memorable 18th of June, the battle of Waterloo. I had seen Comet soon after his birth, and saw him for the last time the week before his death, and although my Duchess 1st, bought at Mr. C. Colling's sale in 1810, was by him, yet I have ever considered Comet the worst blood that has been in that valuable tribe of cattle. For, Mr. C. Colling always said the Duchess cow he bought at Stanwick, was better than any he ever bred from her, although put to his best bulls, which improved all his other tribes of cattle. Even Mr. Maynard's Favourite cow, bred to Hubback, was a better female than she was herself. This Mr. C. Colling has repeatedly acknowledged to me.

Mr. C. Colling never saw my Duchess tribe after his sale in 1810, but I frequently told him that I had improved greatly on Comet by the cross with my Ketton bull (by Favourite, out of Duchess cow, by Daisy), the first Duchess I bought of Mr. C. Colling in 1804. The granddaughter I bought at the sale at Ketton in 1810, at above three times the price I paid for her dam six years before, although she was not so good as her dam. That I have greatly improved on the heifer I bought at the sale in 1810, no one can deny who has seen the 58 Duchess females descended from her, as recorded; but many I did not enter in the Herd Book, when there were no descendants from them. In nothing, however, do they more exceed the original, than in the less consumption of food; while there is a considerably greater growth of carcase, and also a larger supply of butter, though they give less milk than my first Duchess, bought in 1804.

MEMOIR OF MR. BATES,

WITH HIS PERSONAL EXPERIENCE AND RECOLLECTIONS DURING HIS PURSUIT OF AGRICULTURE, AND MORE ESPECIALLY IN REGARD TO HIS PURE HERD OF SHORT-HORNS.

PART I.

Scientific Breeding; a Newspaper Critique; what does it all mean? or what is it all about?—Mr. Bates's Early Life; where and when Born; from whom Descended; had Feeble Health in Early Years; where Educated—Mr. Bates, when young, elected to enter the Church, but his father, though a good Churchman himself, had (for reasons given) great repugnance to his son's wish—Mr. Bates's Schoolfellows and Fellow-Students, and with whom he formed a life-long Friendship—The Blayney Family—The Family Estate left to a Stranger in blood.

"On Wednesday, there was sold at Willis's Rooms a collection of articles, first-rate productions in their way, and the result of extreme industry and skill. Twelve sold for £6,510—*i.e.*, for an average of £542 10s. each. Five sold for £1,699 10s.—*i.e.*, for an average of £339 18s. each. The average of the whole 17 was £481 3s. each. What could these be? First-class pictures? Works of "high art?" Mosaics? Manuscripts? They lived and breathed! They were short-horned cattle. The twelve were cows, and the five bulls. They were animals of a noble race. The catalogue is like a bit from the "Peerage," giving the pedigree of each Grand Duchess or Grand Duke for a dozen or more generations. Mr. Betts, of Preston Hall, Kent, has secured the whole herd of "Grand Duchesses." The "Grand Dukes" are separated; the grandest of all passing, for 600 guineas, to the Duke of Devonshire. The splendour of such an event almost pales the strongest blaze that can be got up by agricultural societies. There is no such test of value, no such triumph of enterprize in that which is obtained, without shows, and judges, and prizes, in the auction room. Here is a plain commercial proof of what can be done, and how far we have advanced upon our forefathers in the matter of kine. But it also proves the difficulty of the work; the necessity of science, and the need of agriculturists educated for their profession. One is almost tempted to ask whether, with such splendid inducements, and such a reality of success, anything remains to be done by societies. But the truth is, it is the breeders of these magnificent animals which must invite every aid which can raise the science of husbandry. Associations have confessedly done much, and are doing much, though apt to try more than they do. They create the taste and skill which appreciate their useful prodigies. They have raised the farmer, who selects the right breed and gives it the right food. Just now, such a society,* one of the oldest in the kingdom, is holding its anniversary, with very little pomp or show, in the province of one of our favourite breeds, in the city of Hereford, and, even by the side of the real work at Willis's rooms, its quiet routine is not altogether uninteresting, &c."†

* The Bath and West of England.

† Leading article in *the Times* newspaper, Friday, 9th June, 1865. The remainder of the article is very instructive on the state of agricultural education, &c.

Any reference to the early life of Mr. Bates, or to his private family affairs, may appear to have no connexion with the subject of the history of improved short-horns; but it seems very difficult to arrive at a due appreciation of the manner in which his herd of improved short-horn cattle was created and maintained in its merits, without, at the same time, considering briefly the various positions in which Mr. Bates was placed, and the talent, education, and study that he devoted to produce the results, as chronicled by the above article in the *Times*. To do this will no doubt be more especially interesting as rural affairs, and the merits of live stock, were not with Mr. Bates an early or natural taste. To understand what he accomplished in perpetuating, if not in perfecting, a race of cattle, the mind cannot but frequently recur to the Author of the work, and therefore I may be excused, if excuse were needed, for referring to his early years and education, and, also, in a cursory way, to the state of agriculture in his native district, and to the connexions and friendships with which he started in life, when he had once devoted himself to agricultural pursuits. A Memoir, with a Portrait, of Mr. Bates, appeared in the Farmers' Magazine, for January, 1850, in which there was given an interesting and accurate account of many parts of his life, and exertions in agricultural improvements. To this I shall, however, refer hereafter.

Mr. Bates was born in Northumberland, in 1775. His father's family had long been settled in the valley of the River Tyne, or, as it is called by the inhabitants, in Tyneside, at Ovington Hall. Thomas Bates, a younger son of the family, for many years represented Morpeth in

Parliament, in the reigns of Queens Mary and Elizabeth, he having been the first M.P. for that borough. He was also at that time supervisor of the royal estates in Northumberland, and the lands of the dissolved ecclesiastical bodies; and, as a reward for his zeal and ability, he obtained grants from the crown of estates, which have continued in the possession of his descendants. The family held considerable property at Horsley, and elsewhere, under the Earls of Northumberland, by the military tenures; but the services and uncertain feudal payments became so oppressive that they were induced, shortly before military tenures were abolished, in the reign of Charles II., to accept leases, which at first were for lives, and then for years, and, finally, under the Duke of Somerset, who had obtained by marriage the Northumberland estates, the lands were let to the highest bidders, by proposals or offers in writing,* and then John Bates, the grandfather of Mr. Bates, left the estate, and lived at Aydon, near Corbridge, an estate which he inherited from his maternal grandfather, John Cook, one of the family of Cook, of Blakemoor, Northumberland. Henry Cook, the uncle of John Bates, was fellow and tutor of Christ's College,

* Since the above was written, my attention has been called by the proceedings of Mr. Stuart Mill and land law reformers, to the operation of this Act of Parliament on the landed property in the kingdom. The militia for the county of Northumberland was, until 1769, raised by the landowners, and the introduction of the ballot occasioned the Hexham riots in that year, when upwards of eighty were killed and wounded, and an officer and several soldiers of the North York militia. The workpeople did not object to the militia, but they objected to compulsory service in it by ballot. Only one person refused to accept leases, and his lands have continued in the possession of his descendants.

Cambridge. He had a promise from the unfortunate Earl of Derwentwater, of Dilston, who was his father's friend and neighbour, of the presentation to the rectory of Simonburn, in North Tynedale, a valuable living extending for above thirty miles. The litigation respecting the tythes in this parish, with the celebrated Dr. Scott, is not yet forgotten in the district. The law was clearly in favour of the rector for all the tythes he claimed, but the inhabitants could not understand why they should pay tythes which had never been paid; and they even burnt the rector's tythe barn, and the produce that he had taken in kind. Happily, this subject of dispute in parishes has been put an end to by legislation, although this was not done till nearly fifty years after the commutation had been urged by George Culley and John Bailey in their Reports, and afterwards proposed by Mr. Pitt, when prime minister.

The Rev. Henry Cook attracted many students from Northumberland to Christ's College, and for many years it was a favourite college in the Northern Counties. Dr. Barker, a relative of the Cook family, was for many years the master. Mrs. Barker was grandmother of Mr. J. Moore Bates, whose son, the Rev. William Bates, D.D., became fellow and tutor, and succeeded Mr. Bates's friend, the Rev. Bernard Gilpin, in the college living of Burnham Westgate, in Norfolk.

Mr. John Bates, to judge from his library, must have had a superior classical and mathematical education, and at one period of his life was extensively engaged in agriculture on his own estate, and the estate of Thornborough, near Corbridge, which belonged to the Earl of Derwentwater, and passed in his attainder to the Commissioners of the Greenwich Hospital. Mr. John Bates was much consulted by the government officers who were sent to take possession of and manage the estate, which was finally settled on the Greenwich Hospital. As an acknowledgement for his services, he was offered a lease of the Dilston estate for his eldest son, George Bates, who, having been educated at the Grammar School of Newcastle-on-Tyne, then in great repute, could not be prevailed on to follow the professional life for which he was intended, but devoted himself to agricultural pursuits. Mr. Geo. Bates married in 1769, at St. Chads, Shrewsbury, Diana, the youngest daughter of Thomas Moore, or More, of the More, in Shropshire, by his wife Ann, youngest daughter of Henry Blayney, Esq.,* of Gregynog in Montgomeryshire. John Moore Bates was the eldest son of the marriage.

Arthur Blayney, the grandson of Henry, the proprietor of the ancient estate, and his sisters were unmarried, and advanced in years; while, of their numerous cousins, two only were married, one to Mr. Thomas, of Aston, in Shropshire, who had only one daughter, after-

---

* Henry Blayney was the son of Sir Arthur Blayney, and Joyous, daughter of John Blayney, one of the Knights of the intended order of the Royal Oak as a reward for their loyalty to the crown; Sir Arthur was second son of Edward, first Lord Blayney by Ann, daughter of Adam Loftus, Archbishop of Dublin, and Lord Chancellor of Ireland, and the founder and first Provost of Trinity College, Dublin. Adam Loftus belonged to the ancient family of Loftus, of Swinehead, in Yorkshire, and having distinguished himself at Trinity College, Cambridge, he attracted the notice of Queen Elizabeth, who sent him to Ireland as Chaplain to the Earl of Essex. The Wellesley family are descendants of Ann Loftus, by her first marriage with Sir George Colley, or Cowley, of Edenderry.

wards married to Sir Cholmondley Edwardes, Bart., of Frodesley, Shropshire, and Mrs. Bates. The succession to the estates rested between the descendants of Mrs. Thomas and Mrs. Bates. As there were no relations in Wales in the male line, it was arranged, at the request of Mr. Blayney and his sisters, that, if the second child of Mrs. Bates was a son, he should be called Arthur Blayney. Mr. Bates was this child, and as he was a very weakly child, and not expected to live, he was baptized and received the name of Thomas from parties to whom the above arrangement was unknown. The name could not be altered afterwards, when he was christened. Arthur Blayney was his godfather. This gentleman's portrait hung in the dining room at Kirklevington.

Mr. Bates had very feeble health in his early years, and had the constant care of his aunt, Miss Joyous Moore, formerly of Millichope, in Shropshire, whose fortune he inherited.

The two brothers were sent to the Haydon Bridge School, of which the Rev. Mr. Hall was master. At a time when all journeys were made on horseback, the local schools were of greater importance than at the present day, and Haydon Bridge School for many years had a very great repute, and was attended by the sons of the neighbouring gentlemen. Mr. Hall had been a fellow of St. John's College, Cambridge. His brother, Dr. Hall, was provost of Trinity College, Dublin, and Bishop of Dromore. They had been educated at the Grammar School of Newcastle-on-Tyne. Among other pupils at that time were the sons of Mr. Tweddell, of Threepwood, who were distinguished scholars at Cambridge. The charge was 25 guineas a year for board and tuition, which may now appear a very small sum, but it was more than the charge at Eton at the same period. Mr. Bates very early elected to enter the Church, and it was intended that he should proceed to Eton. Mr. Hall was very anxious that so promising a pupil should proceed to the university, being very confident that he would obtain distinction there. The state of the Church in the North, at that period, was at a very low ebb indeed: the few benficies worth having were held by absentees, and the small livings and curacies were served by persons who, even when of respectable conduct, were so impoverished, that their social condition was really a disgrace to the nation. There was also at that time much litigation respecting the tythes.

Mr. Geo. Bates had many relations in the Church, and was a good churchman himself, but he had a great repugnance to his son entering the Church, and wished him to follow a country life, and the elder son went from Haydon Bridge to Richmond, in Yorkshire, where there was then a celebrated grammar school, under the Rev. Mr. Temple. Here he had for co-pupils, and made lasting friendships with Mr. John Hutton, of Marske, a well-known agriculturist and improver of his large estates, and also the Rev. James Tate, who afterwards became the master of the school, and canon of St. Paul's Cathedral. Mr. Bates went to Witton-le-Wear, in the County of Durham, to a school of which Mr. Farrer was head master. Some of the letters of Mr. Bates, from Witton, to his father are still in existence, and they show that his studies and pursuits were all with reference to the Church. A gentleman, to whom Mr. Bates introduced me, at Darlington, as a school-fellow, told me, however, that Mr. Bates was not like other boys; he never joined in their play, but would sit for hours in the churchyard with a book. He was soon at the head of the school, and contracted a lasting friendship with Mr. Farrer and his family, and also with the Rev. George Newby, the tutor, and afterwards the head master of the school. After leaving Witton, Mr. Bates, who had passed through the usual course of classical and mathematical study, was a student in the university of Edinburgh, where he subsequently studied in 1809-10-11, during the winter season. Mr. Bates also frequently visited, and passed much time at Gregynog, with Mr. Blayney. The pedigree of the Blayney family, as was formerly usual in the old Welsh houses, was painted on the walls of the dining room at Gregynog, but if Mr. Bates then obtained any impressions about family pedigrees, he paid no attention to them in later life. No Welshman, however, could have devoted more attention to pedigree and family than Mr. Bates did to his favourite cattle.

The memory of Mr. Blayney is still revered in the Welsh border, and probably the best account of his character is that given by his friend, Philip Yorke, Esq., in his "Royal Tribes of Wales" (p. 161), much of which was very applicable to Mr. Bates in his personal habits and pursuits. Mr. Yorke says of Arthur Blayney, that "in his temper he was constitutionally warm. What true Welshman is otherwise? His resentments, generally well founded, were consequently strong, and sometimes permanent. He could forgive an injury, but his confidence once forfeited, it was nearly impossible to retrieve."

The resemblance of Mr. Bates to the portrait of Mr.

Blayney was generally remarked. Mr. Blayney died on the 1st October, 1795, at the age of 85 years, being the same day on which Mr. Bakewell died.*

The Misses Blayney left legacies to Mr. Bates, and much surprise was felt that Mr. Blayney, disregarding all his long expressed opinions and feelings, had, by his will, made shortly before his death, left the ancient estates of his family, which were of very large value, to Lord Tracey, a stranger in blood. They afterwards passed, by marriage, to Lord Sudeley. There was no male representative of this ancient family, except the Lord Blayney in Ireland. Mrs. Bates, and her cousin, Sir Cholmondely Edwardes, of Shrewsbury, were the co-heirs. Mr. George Bates could not be prevailed on to dispute the will. Mr. Bates probably felt disappointed, but he never mentioned Mr. Blayney, except as to the coincidence of his death with Mr. Bakewell, and as being his godfather.

---

* The *Gentleman's Magazine*, 1795, page 881, in the obituary of remarkable persons, has the following notice:—" At Gregynog, in Montgomeryshire, in his 81st year (85 really), Arthur Blayney, Esq. This worthy gentleman, for he was very properly styled, 'The father of Montgomeryshire,' was the common friend of the poor and distressed, and his death will be long and deeply lamented in his own neighbourhood, and in the adjoining parishes around his mansion. By his unremitting exertions and most liberal assistance he has given a new face to the surrounding country. His tenantry will have great cause to lament his death; for he has not raised the rent of his farms for more than forty years. The great road, the canal, the church, will be lasting monuments of his perseverance and public spirit. He was buried in Tregynew churchyard, the family vault having been stopped up several years since by his order, from a dislike he had to interment in churches, and by his express desire the funeral was very private." The notice of Mr. Bakewell, is at 971 of the same volume.

PART II.

Mr. Bates's Agricultural Career began at Aydon Castle—The Families of Surtees, Bates, and Cook, and the intimacy of the Culleys with Mr. John Bates and his Sons—Mr. George Culley and Arthur Young—Teeswater Landowners and Occupiers; their character, and Mr. George Bates's intimacy with them—The Messrs. Culley, Mr. George Bates, and Major Surtees—Messrs. Culley's Rent £10,000 a-year—Mr. George Culley's judgment criticised, but not weakened—Mr. Bates a visitor to, but not a pupil of, Mr. George Culley, who, however, ultimately became the model agriculturist whom Mr. Bates followed—Mr. George Culley died in 1817—Mr. Bates began Farming before of age; his earnestness, energy, and successful treatment of different soils—Secret Tenders in Letting Farms; when inaugurated—Tyneside Farmers in 1800 and their Short-horn Cattle—Mr. Jobling, of Styford, Mr. Wastell, Messrs. Colling, and Mr. Donkin, of Prudhoe—What Landowners had the best breeds of Cattle—Mr. Bates enters upon Halton Castle, and his surrounding Influences—Mr. Bates and the "Wonderful Ox," and Mr. Maynard's Steers and Heifers—Kyloes in Northumberland; their handling and fatting nature as a pure breed, and when crossed with Short-horns—Weights of remarkable Oxen—Mr. Harbottle on Mr. Bates's Beef as sent to the Smithfield Club Dinner—Mr. George Culley and Mr. Bailey on Agricultural Societies—Mr. John Grey's (Dilston) Assumptions.

We now come to the beginning of the agricultural career of Mr. Bates, which began with his agricultural education on the farm of Aydon Castle, which adjoined the estate of Aydon White House. For a few years the managment and conducting of the farm was left almost entirely to Mr. Bates. Here some other names, well known in connection with improved agriculture, and agricultural literature, must be introduced. Matthew and Geo. Culley, were the sons of a gentleman of good landed estate at Denton-on-the-Tees, in the County of Durham. Their mother was Eleanor, daughter of Edward Surtees, Esq., of

Mainsforth, in that county, a family well known as that of the learned historian of that county.

There had been several marriages between the families of Surtees, Bates, and Cook, and a great intimacy existed between the Culley family and Mr. John Bates and his sons. Geo. Culley was an intimate friend of Mr. Bakewell, and had spent a year at Dishley. George Culley and Mr. Bakewell were constant correspondents, and frequently made agricultural tours together. Mr. Geo. Culley was a friend and frequent correspondent of Arthur Young, and contributor to the "Annals of Agriculture." Matthew Culley married Elizabeth Bates, a cousin of Mr. Geo. Bates.

The valley of the Tees contained in the last century, a large number of landowners and occupiers who were second to none in the kingdom for their knowledge and practice of rural affairs, and Mr. Geo. Bates had an intimate acquaintance or friendship with the most eminent of them; and all improvements, both in stock-breeding and cultivating the soil, were soon copied or rivalled by an equally educated and intelligent class upon Tyneside, where there was much land specially adapted to the growth of turnips and green crops, the cultivation of which was very early introduced there, while the feeding of stock was much stimulated by the great demand at Newcastle for supplying the shipping and coal works.

The Messrs. Culley were frequent visitors to Mr. George Bates, and Major Surtees,* of Newbiggin, near

---

* This gentleman commanded the Northumberland Militia in suppressing the Lord George Gordon riots in London, and was a correspondent of Arthur Young, and devoted much attention to the improvement of his woodlands.

Hexham, and they there saw the success of the turnip cultivation in Tyneside, which induced them to endeavour to obtain the Dilston estate, which Mr. John Bates had declined. Ultimately, however, they settled, in 1767, at Fenton, in the Tweedside district, and they very soon extended their agricultural occupations, until they paid above £10,000 a year rent. In this business pursuit they added very largely to their fortunes, and when they gave up business, Matthew retired to the Copeland Castle estate, where his descendants now reside, his grandson having served the office of Sheriff for the County of Northumberland, while George Culley retired to Fowberry Tower, and his great grandson was Sheriff.

As a matter of business arrangement between the brothers, George Culley undertook the buying of the stock, and the selling of the produce of their farms, and a very full occupation it must have been. Besides the cattle and sheep which they bred, George Culley yearly purchased at Yarm fair, large droves of oxen for working purposes, of which they usually kept 150, and fatted them off after two or three years' work.

Mr. Culley's judgment has frequently been the subject of captious criticism and denunciation. A modern school of short-horn breeders, indeed, apparently wish to ignore altogether the principle, before mentioned, of *handling*, as laid down by Mr. Culley. They must, however, recollect that Mr. Culley was speaking with the experience of a man in the daily practice of buying and feeding cattle, and, therefore, he was able to test the feeding qualities of his herds. He was not writing to elevate one class of breeders above another, or to serve the purposes of any school of breeders, or the sale of their herds. Mr. Culley in his writings did full justice to the various Scotch breeds, and their value in the London markets, and he states that many polled cattle were good handlers and great feeders.

Mr. Bates was not a pupil of George Culley; but he paid him long visits, and was in constant intercourse with him, and familiar with the state and management of Fenton, Wark, and their other farms, and no doubt profited much by Mr. Culley's instructions. Mr. Culley was, indeed, his model, and the person he looked up to for all agricultural improvements. The only point on which Mr. Bates did not agree with Mr. Culley, was respecting short-horn cows, Mr. Culley having the very usual impression that large milkers or good dairy cows were often not good grazers or feeders, and Mr. Bates's frequent reference to the dairy qualities of his short-horn stock combated the opinion of Mr. Culley. Mr. George Culley died in 1817.

I now return to Mr. Bates. Before he was twenty-one he became tenant of his father, on the estate of Park End in Wark Eeles or Islands, in the vale of North Tyne. The rent was nominal, and the estate consisted partly of a considerable portion of Haugh or Holme land near the river, which had formed islands and was frequently flooded by the river, and partly of some allotments of common. Mr. Bates soon showed his aptitude for farming and improving land, by the progress of the agriculture on the estate. He embanked the haugh from the river, being the first work of that sort on the Tyne. He planted the hill sides, drained all that required it, took up the old fences and planted fresh ones, and collected the water and drove the thrashing machine according to Mr. Bailey's plan in the "Survey of Northumberland." Mr. Bates was engaged here until the year 1800, when the farms of the Halton Castle estate became vacant. Halton Castle had been some years in the occupation of Mr. Thos. Bates, the brother-in-law of Mr. Matthew Culley, with another farm occupied by Messrs. Bell. Mr. Thomas Bates had retired to his estate at Brunton, near Chollerford, North Tyne, which, however, he afterwards sold to Henry Tulip, Esq., and then purchased a more extensive estate at Akeld, near Wooler, adjoining the estate of Copeland Castle. Mr. Thos. Bates was a well known agriculturist, and when at Halton Castle had numerous pupils, especially from the border counties, and among others, Mr. Thos. Scott, of Leitham, near Jedburgh, a gentleman well known and much respected on the Scotish Border, and a very successful stock farmer both on the borders and in the highlands of Scotland.

The practice of letting farms by secret proposals, it may be here remarked, was never a favourite one with agriculturists, and it is reprobated in the "Survey of Northumberland," (page 25). The system was introduced into the North of England by the Greenwich Hospital Commissioners, and it was considered by them to be the proper mode of letting what were, in fact, the public lands. The farms of Halton Castle contained about 800 acres, some of which was valuable old grass land. Much against the wish of his father, who advised him to be satisfied with his own estate of Park End, Mr. Bates offered for the Halton Castle estate, and was the highest bidder by about £125 a year, for a 21 years' lease. He now had ample scope for his exertions, and threw himself with energy into his work.

He at once drained the greater part of it on the Elkington principle, with drains that were often ten feet deep. It may be here questioned, whether the practice of draining has been improved by the modern usage of putting in drains at a uniform depth, generally of four feet, and at a uniform distance, without much reference to the nature of the land, or the sources of the water. The Elkington principle, for which the Government granted £10,000, was to carefully study the nature of the soil and strata, and to cut off by drains the springs and flow of water. The drains were often more expensive than the modern tile drains, but then every drain was of use. Whereas in the "thorough" draining, a large per centage of the pipes very often never contained water at all. Lime, which was abundant on the estate, was very extensively used after draining.

In 1800 the agriculturists in Tyneside, and other parts of Northumberland, were amongst the most eminent in the kingdom, and nowhere had the breeds of stock been more improved, or more numerous herds exhibited. The first volumes of the Herd Book shows that most of the best animals had been located in Northumberland. The Leicester sheep, introduced by George Culley and his friends, had superseded the old "Mugs," and were kept a pure breed. The practice of crossing with the Cheviot had not been introduced.

Mr. Thomas Jobling, of Styford, and other members of his family, had long been well known as agriculturists, and had early obtained bulls from Mr. Wastell, and others, on the Tees, and afterwards from the Messrs. Colling, and had a breed which was long considered one of the best in the North, and Messrs. Colling often obtained animals from him. Bolingbroke (86), the sire of Favourite (252), was long stationed at Styford, and his stock being known by their white faces and red bodies, much resembling the Herefords in colour. Styford and Daisy bulls had also been used by Mr. Thomas Jobling.

Mr. Donkin, of Sandhoe, near Hexham, had also a good herd. His uncles, as stated by Arthur Young, introduced the drill system for turnips, having practised it in Yorkshire. The Angus family, at Hindley, were the owners of a good herd. Geo. Gibson, Esq., of Stagshaw Close House, Sir William Loraine, Bart., of Kirk Harle, and numerous other landowners and occupiers, were possessed of the best improved breeds of cattle and sheep. The Commissioners of Greenwich Hospital were the owners of many of the best farms in Tyneside, and under the able and liberal management of Mr. Lockyer, the secretary, were foremost in all the improvements of the district. Their income from land and mines was above £130,000: and chiefly by Mr. Lockyer's exertions, the turnpike roads along both sides of the river Tyne, and to Alston, were made. The Greenwich Hospital, in common with other landowners, felt the impulse of the large increase of rents, from improved cultivation and stock, and the French War; and on their estates and nearly all others, the old farm houses and buildings were soon replaced by extensive and often massive stone buildings covered with blue slates. The Greenwich Hospital estate probably suffered from the want of a resident owner. The mining income was so large that the landed estate was often a secondary matter, the smaller tenants were chiefly employed in conveying the lead from the mines to the shipping places. The head agents or commissioners generally were not conversant with estate management. Smeaton, the engineer of the Eddystone lighthouse, and also the great tunnel in the lead mines, was long a commissioner. But this defect in the management of the landed property was sought to be remedied by having local agents or bailiffs, who were selected from the leading agriculturists of the district. Mr. Wm. Jobling, of Newton Hall, whose son, Cresswell Jobling, was many years Chairman of the Quarter Sessions of the county, was bailiff for the Corbridge district, and was succeeded by Mr. Anthony Wailes. Mr. Coates, of Lipwood, and Mr. Sample, were the bailiffs for the Haydon Bridge and South Tyne districts, and Mr. Thos. Ridley for the North Tyne and hill districts. These gentlemen were well known agriculturists, and also considerable landowners, and the other districts had bailiffs of the same class.

Mr. Bates, on entering upon Halton Castle estate, had plenty of good agriculture to copy from, and compete with. He obtained many excellent cattle from his father; and also the best class of the improved Leicester sheep. He also obtained some Cleveland bay horses, which at that period had attained great perfection on Tyneside.* The swine, and even poultry, did not escape his attention, but it was

* This most valuable breed of horses is, I believe, now quite extinct in Northumberland. They were all bays, and the old breed had generally white faces, and white legs, but, afterwards they were pure bay, without white. No sire was considered pure bred that produced a foal not all bay, whatever might be the colour or breed of the dam. The last, I believe, of the breed were taken by Mr. Edward Bates to Germany.

to his herd of cattle that he devoted his greatest attention. He bought cows of Messrs. Colling in 1800, but I can find no record of them. On his Park End farm, in North Tyne, he had established a herd principally selected from his father's stock; and he purchased the Daisy bull (186), from Mr. Thomas Jobling, for 30 guineas, and in 1804, he bought the first Duchess, as before mentioned. I have never heard the particulars respecting the purchase of her daughter, but no doubt Mr. Bates purchased her, and Mrs. C. Colling's surprise at obtaining 100 guineas for the cow was so great, that she immediately suspected they must be worth far more money, and wished to cancel the sale, especially, when she found that Mr. Bates attached so much importance to the Duchess blood. It may be remarked that this lady took a leading part in the management of the herd.

R. & C. Culley were the sons of a farmer, near Darlington, in the County of Durham, and who do not appear to have been educated beyond the ordinary rules. Robert, it is said, had originally been intended for trade, Charles was, however, a man of superior intelligence, and mixed in good society, and was usually looked up to in the County. He had also been a pupil and friend of Mr. Bakewell. They had received great attention and kindness from George Culley. Their portraits were painted on the same picture, and certainly do not give the idea of very intellectual men. Mr. Bates, when very young, became a frequent visitor at their farms, and knew every animal in their herd, from his connexion with the Messrs. Culley. Messrs. Colling were always most kind and attentive. Mr. Bates, when very young, generally attended Darlington great markets, always held on Mondays, and he generally visited from the Saturday night to Monday with Messrs. Collings, or Mr. Mason at Chilton. Mr. Bates had seen the travelling (or Wonderful) Durham ox, bred by Mr. C. Colling, and the white heifer of Mr. R. Colling, so often mentioned in accounts of short-horns, but the roan heifer of the Duchess blood attracted the attention of Mr. Bates more than the other two, and this, no doubt, made him resolve to have that blood if he could obtain it. It was at the great Darlington market, on the 1st Monday in March, 1799, that the attention of Mr. Bates was drawn to this heifer. This market was attended by all the leading country gentlemen and agriculturists, and the winter fed cattle where then exhibited for sale, and great rivalry existed in the display of cattle. I have often heard of the annual display of the stock of Mr. Maynard, of Eryholme who generally exhibited 8 steers, and 8 heifers, opposite the King's Head Inn. The wonderful ox was the great sight in this year. But Mr. Bates has recorded in his writings how he thrice met Mr. Thompson, a well known judge of stock from Northumberland, by the roan heifer.

The West Highland cattle or kyloes were grazed in large numbers in Northumberland, and every year the Messrs. Magnay, Armstrong, and other extensive dealers, took numerous droves through the county to the South of England. Mr. Moorhouse, who lived near Skipton, in Yorkshire, also imported largely; and under the direction of Mr. Bates he purchased heifers from Mr. McDougal, of Lorne, and other well known gentlemen in the West Highlands, who had paid great attention to their native breed.

The hair, handling and aptitude to fatten of the best

class of West Highland cattle is well known, but Mr. Bates was the first person to try the cross between the best short-horn bulls and the best West Highlanders, which he called an improved breed of cattle; and I have been informed that he entertained the plan of establishing a new breed of cattle, to be perpetuated by breeding from the cross bred cattle. But he never kept any cross bred bulls. Many of these cross bred cattle obtained a great size, with great aptitude to fatten. Two remarkable oxen of this class were fed by him and exhibited, in February, 1808, at Newcastle-on-Tyne, for the benefit of the Infirmary, and were slaughtered in that town. The weight of one of these was, fore quarters, 51st. 13lbs.; hind quarters, 45st. 2lbs.; tallow, 14st. 10lbs.; hide, 6st. 12lbs.—118st. 9lbs., of 14lbs. to the stone, and it was sold by Mr. Wm. Robson, the chief butcher in the town, for £75 7s. 5d. The portraits of these oxen, one white, and the other brindled, hung in the dining room at Kirklevington. The prints were for many years to be found among those of celebrated cattle, and at the foot were the particulars of the dimensions of them. This was then considered a new era in cattle breeding, as it was the first attempt at cross breeding, and the success attracted the attention of the agricultural societies in all parts of the kingdom.

Mr. George Harbottle, who had left Anick Grainge, near Hexham, for a large farm near Henley-on-Thames, and who afterwards went to Russia as agent to Prince Demidoff, relates in the following letter the attention paid to this class of breeding by the original Smithfield Club, of which Lord Somerville, a well known agriculturist, was president:—

"London, 2nd March, 1808.—My Dear Sir,—Having had the honour of a card to Lord Somerville's Annual Dinner, which was held yesterday, at the Freemason's Tavern, I was agreeably surprised to hear from his Lordship, that a piece of your fat ox had been on the table; and persuming you would be gratified on hearing the reception it met with, I am induced to address a few lines to you on the occasion. After disposing of the prizes, his Lordship begged leave to inform the company that a piece of beef, of the crossed breed (describing it) had that day been on the table. The gentleman, he said, he had not the pleasure of knowing, but it was obligingly sent by a Mr. Bates, of Northumberland, remarking as he had said before he had not the pleasure of knowing him, yet, he trusted the gentlemen present would join him in drinking Mr. Bates's health. This was received with unusual approbation, and without flattering, I assure you I was not the most silent with my knuckles on the table on the cccasion. The beef was not on the table where I sat, consequently I am unable to tell you how it ate. The party to dinner amounted to about 400, Lord Somerville in the chair, and at his right hand was the Duke of Clarence, and amongst others I noticed the Duke of Bedford, Sir J. Sinclair, Sir Thomas Seabright, Sir Thomas Carr, &c., &c. These gentlefolks were at a cross table at the head of the room, and the rest of the company were disposed in four tables down the room. The show was inferior to what I have seen. The Duke of Bedford won the 1st prize for the best Devon ox. The names of the other gentlemen, who won the other prizes, I do not recollect, except the Earl of Bridgewater, who has a silver cup for the best pen of five Southdown doe hoggs. Sir H. Vane showed a very good Highland stot. Amongst the other beasts there was nothing particular, except the carcase of a three shear Leicester wether, which weighed 185lbs."

Agricultural societies, for various purposes and objects, have, no doubt, long existed. George Culley, in writing of them, in 1797, in the "Survey of Northumberland," says, "There never was any agricultural society in this county; and, if any ever had existed, it probably would have been soon dissolved, if we may judge from the experiments that have been made in some neighbouring districts, where we find that after a few years' continuance, they have been given up; but whether from a radical defect in the institutions, the non-attendance and indifference of members, or the *injudicious distribution of prizes*, we are not prepared to say; but think that public farms are much more likely to promote improvements in the science of agriculture."

Mr. Bailey, in his "Survey of Durham," enumerates the agricultural societies in the north. The County of Durham Society has been referred to, and the Rusheyford Society, near where Mr. Mason lived, had, perhaps, rules more likely to be of permanent value than any other. The Tyneside Society was established in 1804, and Mr. Bates took an active part in promoting its success. He was a large exhibitor at the shows. The bulls were shown in the spring, and Mr. Bates generally obtained the prize, but, on account of the conditions as to the use of the bulls, he did not show his best. I find him, in one year, obtaining the prize for the best bull, for the best crop of turnips, beating Sir William Loraine; and also the prize for the best cow, best sow, best road mare, and best sheep.*

PART III.

MR. BATES'S ADDRESS TO THE BOARD OF AGRICULTURE.

Mr. Bates was not satisfied with the proceedings of the Tyneside society stopping short in merely giving prizes. He wished to follow out the views and principles of George Culley, in regard to experiments and trials to be made, proposals for testing the value of divers methods of cultivation, and the value of stock in feeding. With his usual determination when he had come to a conclusion, Mr. Bates, on finding the society would not accede to his wishes, embodied his views in a letter, which he addressed to the Board of Agriculture, and Agricultural Societies. I venture to put this production in here, as showing his views at that period:—

Address to the Board of Agriculture, and to the other Agricultural Societies of the Kingdom, on the Importance of an Institution for ascertaining the Merits of the different Breeds of Live Stock; pointing out the advantages that will accrue therefrom to the Landed Interest, and the Kingdom in general, by Thomas Bates.

---

* A recent work, "the life of John Grey, of Dilston," assumes that agricultural societies were unknown in the North of England until introduced by Mr. Grey, after 1830. The meeting of a new society in Tyneside, in 1837, was attended by Mr. Bates, and also by Mr. Charge, of Barton, and other gentlemen from the Tees. The observations of Mr. Bates appear in his statements. Mr. Charge enquired for the well known names at the former meeting, which he had attended at Ovingham. They had nearly all disappeared; and Mr. Charge remarked that the agriculturists and the stock had equally degenerated.

The *past* is of all others, the most valuable source of instruction for the *future*. The lessons it teaches are often dearly purchased; but nations, as well as individuals, have borne ample testimony to the salutary effects that must always result from an attention to them.

Let us consider only the Agricultural History of this country in our own day. 'Till very lately our manufactures and commerce were looked upon as almost the sole sources of our national wealth, They were, accordingly, almost the only objects of legislative attention and favour. Under these discouraging cirumstances, agriculture, except in a very few illustrious instances, was practised only by the illiterate and unenlightened, who with small capitals, with little inclination, and with still less knowledge for prosecuting such enquiries as are always the harbingers of improvements, plodded on most diligently, without deviation from the path of their forefathers. Thus, uncherished by the protecting hand of power, and the intelligence and capital of the nation in a great measure directed to other objects, agriculture languished, or at least made inconsiderable advances towards improvement.

But from an apathy so fatal to its best and most substantial interests, the country was roused by years of scarcity. It was at length discovered that some little attention was necessary to be paid to the interests of agriculture. Foremost in the list of those who directed the efforts of the legislature to this desirable purpose, stood a meritorious and most valuable member of society (Sir John Sinclair), whose patriotic zeal in the cause is entitled to the highest praise. The establishment of the Board of Agriculture; the consequent survey of every county in the kingdom, which afforded the most rapid and efficacious means of conveying the improvements of one district to another; the enthusiasm for the profession imparted to all ranks from the nobleman to the mechanic, which threw into it a large additional capital; the formation of provincial societies, which are so well calculated to diffuse and keep alive the spirit of investigation and to encourage research: all these are the effects of the national genius being directed towards agriculture, and have already produced a return of almost incalculable benefits. It is a fact worth whole volumes of abstract reasoning, and shews how well improvements have been repaid where encouragement has been given to agriculture, that some farms in the district from whence this is written, pay now a greater annual sum in direct taxes on the land than the amount of their rentals during the American War; and yet, from the improved cultivation of their soil, and the still greater superiority of the live-stock kept upon them, the present occupiers realize much greater returns for their skill and capital, than were obtained at that period. Does not this account in a great measure, for the depression the country felt at that period, when taxes were trivial in comparison to the present times? Had agriculture not been improved since then, how deplorable would have been our present situation, or would it have been possible for us to maintain the arduous contest in which we are now engaged?

Much has been done in some districts, but a wide field for improvement still lies before us, and if ever there was a moment in the History of Britain, which called upon her sons for every kind of exertion, it is the present. In order, therefore, that we may face with confidence, the unparalleled dangers to which we are exposed, ought we not to have recourse to unparalleled exertions? Cut off from all foreign aid in the time of need, ought we not to cultivate with increased diligence our own resources, and improve and extend them to the utmost? Our extensive colonial possessions will always ensure us a supply of all the usual articles of luxury, but in such a situation, ought we not to enquire with more anxiety and earnestness than ever, how far our country has hitherto been able to furnish from its own soil, that absolute necessary of life, *Bread for its inhabitants?* Now, it is ascertained that Great Britain has, in the best of years, supplied barely a sufficiency for its population, and when, from unfavourable seasons, its produce has been diminished, we have been forced to depend on a foreign supply. Thus, it becomes a question of most serious urgency, what means can be devised to obviate the dangers of such a situation?

To this great end, I conceive that nothing can contribute more than the skilful selection and improvement of the breeds of live stock. For there is no doubt that the various breeds differ very materially, both in the length of time, and in the quantity of food which are requisite to fit them for slaughter. Hence, by choosing the proper species of stock, a part of those extensive tracts of land which are at present set apart as pasturage for rearing and fattening them, may be taken into cultivation, and by adopting alternate rotations of corn and green crops, we may raise a greater quantity of grain, while, at the same time, as great a proportion of live stock may still be maintained. This is particularly desirable in our present situation, as we must raise many other articles of unavoidable necessity (and less land of course will be applied to the growth of grain), particularly hemp, flax, &c., which we have heretofore been accustomed to import.

But, in order that this system may be perfected, and brought into general use, in order that the *best* varieties of each species of stock may be universally adopted (a circumstance on which the value of the green crops and pasturage of the kingdom depend), it is necessary that accurate experiments should be made to ascertain the comparative merit of each variety, with a fair statement of the expenditure and return. Were all farmers made sensible, by such experiments, of the real difference which there is in stock, their own interest would prompt them to adopt the best, but while they remain ignorant thereof, they cannot be expected to commence improvements; and what will so soon convince them of this difference, as accurate experiments, on the impartiality of which they can rely? And where is impartiality to be looked for, if it cannot be found in a national institution, where there can be no interest in any deception? By these means, it would then be known in what consisted the superiority of one kind of stock over another; whereas, at present, such is the want of knowledge on the subject, that cautious persons choose rather to keep their money in their pockets, than expend it for an uncertainty. And perhaps many men of skill in this line are little known, because their modesty hides that merit which pushing, enterprising men would bring into view. And every one knows how far a high name goes in disposing of stock that is called "well descended."

The improvement of the live stock would not require one additional hand to be employed; and, if but one pound of butcher's meat more in ten could be obtained (and those who are well conversant in the feeding of stock will readily admit, from the difference which they have already found in what has passed through their hands, that more than double this increase might be expected), how much would it contribute to the advantage of the farming interest? and, if the live stock were universally improved in this proportion, how great an addition would be made to the prosperity of the nation?

It is true that the breeders, by whom the improvement was made, would reap the benefit in the first instance, by the higher prices which they would receive for their stock. But the purchasers or hirers would also be eventually recompensed for the extraordinary prices they paid, by the increase in the value of the produce of their stock. Thus, while the purchaser or hirer received a fair return for his money, the eminent breeder, by his rising profits, would be encouraged to further exertions; and others would be stimulated by his success to try what they also could accomplish in the way of improvement. Hence would arise an inexhaustible source of wealth to the nation; as may be proved by the districts where attention has been paid to this subject, within the last thirty years. And what is now wished is to awaken and stimulate those districts to similar exertions which have not hitherto attended to it, or, at least, in a very partial way.

The high importance of this subject is evident from the observations of those who are more immediately interested in attending to it, as all those who have made the trial acknowledge that they find a great and material difference in the variety of breeds, not only in the quickness of thriving, but also in the consumption of food. It is only of late, however, that any minute attention has been paid to the selection of breeding stock. Indeed, from any records that are now extant, we find no person who was eminent in these investigations until the time of Mr. Bakewell, who, having shewn how much might be done by such trials, has excited breeders, throughout many parts of the kingdom, to make similar observations, either upon the stock of the district where they resided, or upon such varieties as they brought from other districts, where judgment or fancy led them to expect a superiority: and agricultural societies have been established in almost every district, to forward this great object amongst others.

K

But when a comprehensive mind views what has been done, and considers the present state of the different species of live stock throughout the kingdom, what information can he gather from all the observations hitherto recorded, to regulate his selection of any one variety as more suitable than another, for any particular purpose he may have in view? Excepting the race-horse, whose speed and bottom are pretty accurately known, we have made but few *precise* enquiries into the merits of other animals, although many of them are perhaps more deserving of our attention. We have, indeed, by experience discovered in general, that there is a material difference in stock, but what, or how great that differance is, we have not yet been informed with any precision or certainty.

This important information I fear, has not hitherto been much advanced, in consequence of the premiums offered by the agricultural societies, from the respective merits of the different breeds of stock not being sufficiently ascertained. It is true that such persons as are supposed to possess the greatest knowledge are generally appointed judges in awarding the premiums; but what one set of judges think right one year, another often think wrong the next, or, which is generally the case, they determine that stock to be the best to which they have been most accustomed, or which they possess themselves. The long or the short-horned are favoured by those who breed the one or the other species, and so in other cases, and no discovery is made of the different merits in the stock exhibited, but only of the prejudices, or at best of the sentiments and opinions of those who act as judges. In fact many individuals are discouraged from introducing breeds different from those which are prevalent in the district, on account of the difficulties, hazard, and prejudices, which they have to surmount.

A striking instance of the influence of prejudice occurred when Messrs. Culley and Thompson first introduced the Leicestershire, or Dishley breed of sheep, into the northern part of this county. The dealers from Yorkshire, who bought the *cart stock* of that district in the autumn, refused to purchase this breed, and Messrs. Culley and Thompson were obliged to feed them on turnips in their own neighbourhood, until the Yorkshire dealers were convinced that the sheep which they had rejected were quicker thrivers and more profitable than the breed of the district. And so great has been the change since the prejudice was removed, that scarcely a vestige now remains of the original stock.

Some societies determine the comparative merit of animals to be slaughtered, by weighing them alive, and their carcases when dead, and taking the smallness of the offal as the criterion by which the best may be distinguished. But no accurate judgment can be drawn from hence, since *poor keep* when young greatly tends to the increase of offal (tallow excepted), for the smallness of the intestines depends much on the quality of the animal's food when young, as well as on other causes. Other modes, it is true, are adopted by the more enlightened judges, such as fineness of bone, true symmetry of form, quality of the flesh, which is ascertained by being peculiarly mellow and free to the touch, indicating a strong tendancy or disposition to fatten, &c. But, in many districts these circumstances are not attended to; and as the looks and handling of animals depend greatly upon their keep, and the degree of fatness at the time, nothing can be determined with precision, unless the stock has been for some time kept on the same food.

The premiums also of such societies, even under the best regulations, often go into the hands of only a few individuals, which discourages those who see no prospect of succeeding; and in a few years they are generally either much neglected, or wholly discontinued. Besides, the expense and trouble to which individuals are put in the keep of their stock, in order to gain the premiums, exceeds sometimes the value to be obtained.

There is one great good, however, which has resulted from these societies, and that is the creating and diffusing of that *spirit* for improvement and free enquiry, which appears, at present, almost universally to prevail, even in the remotest parts of the Highlands of Scotland. In a tour which I made through that district, two years ago, I found this spirit as equally alive as in the more favoured counties. But there, as elsewhere, the farmers appeared greatly at a loss which species to select, and were breeding from the coarsest and largest boned animals, because they found them most agreeable to the Yorkshire dealers, from an absurd emulation among the graziers there, who should sell his cattle at the greatest price per head, without attending to the cost which he incurred in making the animal so high-priced. From this cause, the Highlanders have been for some years losing by neglect the valuable qualities which their stock possess. This is the more to be wondered at in a people so well informed as the Highlanders, and who shew such quickness of conception on other subjects.

It behoves, therefore, the well-wishers to the cause to keep up that spirit, and *institute experiments*, which should be carried on accurately in different districts, under every variety of climate, soil, and situation, in order to determine the *merits* of each species of live stock that may be thought worthy of the trial, by ascertaining the quantity of food consumed, as well as the improvement and return produced.

The great importance of a general institution for carrying on experiments, according to this plan, is what I am anxious to submit to the consideration of the Board of Agriculture, and of all the Agricultural Societies. Now, it is evident that by this means the intrinsic worth of every particular breed, and its comparative value with other breeds suitable for the same purposes, would be accurately known. In this way, the best breed for any particular purpose would at last be discovered—a discovery which would be universally beneficial. Even the plain common farmer would know the real merits of the stock in his possession, and adopt what was best. The individuals possessed of the best breeds would receive in the prices of what they disposed of, or let out, the sure and true reward of their merits, together with the high gratification of having rendered an important service to their country. The purchasers, or hirers, would be sensible that they paid no more than an adequate price. The whole farming interest would be well rewarded for their exertions; and, as they could then afford to pay a greater rent, the landed proprietors would be particularly interested in the result; for all agricultural improvements eventually tend to their benefit. At the same time, the public would be furnished with an additional quantity of food, which would support an increasing population, or, at least, be sufficient for our present numbers, without dependence upon foreign aid. And if, as an eminent writer has observed, "he who causes two blades of grass to grow, where only one grew before, is deserving better of his country than all the race of politicians put together;" surely they are not less deserving who raise a greater proportion of animal food, from a given quantity of vegetable production than has been hitherto done.

One very great evil, which I have not yet mentioned, would be prevented by such an institution as I am now endeavouring to recommend: it is this; those breeders whose stock is in general estimation, and who are in the habit of letting out males by the season, or who breed with an intention of disposing of their females, keep them very high while young, and use all methods of forcing them forward in condition, and many who wish to have their stock in repute keep all their breeding stock high, a practice which evidently tends to weaken the constitution of the animals themselves, which are thus treated, and, of course, to debilitate their offspring. From this cause, more perhaps than any other, we find breeders complaining that they are disappointed in the produce of their stock; and, indeed, many animals that are called *high-bred*, produce very puny, weak, and tender stock, and are not prolific. Now, if the merit of every breed were ascertained there would be no necessity for such a practice, an enormous expense would also be saved to the breeders, and, moreover, the public would derive a benefit, since the keep given to those forced animals would be applied to feed off such as are intended for slaughter, and thus to produce a greater quantity of animal food. It would also prevent or destroy those combinations which sometimes take place amongst the leading breeders of a fashionable or favourite species of stock.

The more the subject is investigated, the greater becomes our conviction of the extent of improvement of which it is susceptible. But improvement in this department, is not so quickly accomplished as in cultivation, where a more beneficial mode of practice than what we knew before is easily seen, and readily adopted; besides that, a new or improved variety of grain, or seed, when once selected, is soon diffused, and multiplies much faster, as well as much more than animals. This consideration, however, should only excite us to lose no time in setting about our plans for their improvement.

It is true, that some agriculturists have been trying experiments on these subjects, for their own information; and the public are highly indebted to a few distinguished patriots, who have communicated their experiments for general benefit. Among the foremost of these I cannot help mentioning Mr. Curwen, whose liberal communications of the experiments which he carries on at Workington, as well as the encouragement which he gives to the exertions of others, are highly honourable to his name, and must render the society over which he presides, the resort of genius, and of distinguished agriculturists from all parts of the United Kingdom. The Duke of Bedford, I am informed, also pursues with much zeal and intelligence, the various plans of agricultural improvement, carried to so great an extent by his ever to be lamented brother. Lord Somerville's exertions are also well known to the public. But till these experiments are carried on with different species of stock, and upon a much more extensive plan than a few individuals, however eminent, are competent to accomplish, they cannot be productive of accurate results. Whereas, if a public institution of the nature now recommended were once established, there is every reason to believe that it would create a new and important era in agriculture. It would ascertain the value of each species of food, for the purposes of producing beef and mutton, as well as for the dairy, &c., &c., and a closer attention to the succeeding rotation of crops, would further tend to appreciate their use.

I can see no objections to this plan, except such as may be made by individuals, who are in possession of breeds, from which they derive, at psesent, great emoluments, and who might be apprehensive of their diminution, if the comparative merits of different breeds were accurately ascertained. These persons may be cool at the first in adopting such a plan, and in allowing their stock to be brought to the test. Yet, I would gladly hope that they possess public spirit sufficient to run the risk of sacrificing their private advantages for the general good, since in case of success to their stock, it would establish their reputation without any future dread of disparagement, and consequently increase the demand. Moreover, as the refusal of those who possess the best varieties of any breed, will render that breed liable to disrepute, from the experiments being tried upon the inferior varieties, they would readily see the risk to which they are exposed, by not acceding to this plan, if the landed proprietors enter into it with that spirit which their own interest (in raising thereby the value of their proprety), and the public benefit so evidently require.

Those breeds too, which, for any length of time have been attended to, have evidently an advantage over those which are at present neglected, or have but lately been brought into notice; and if the persons possessing the former, hesitate in such a case, little doubt will remain to what cause their conduct is to be attributed.

It deserves also to be suggested that the longer this desirable institution is delayed, the greater will be the number of prejudiced individuals; for the possessors of a breed which happens to be in repute, by putting other persons in possession of a few of their stock, will make them partakers of its advantages, and thus link them to their cause, and render them instrumental in preventing an accurate investigation into the merits of the breed.

To draw up any plan in detail, in the present state of the discussion, appears to me premature and unnecessary, until it be known what encouragement will be afforded by the Board of Agriculture and the other Agricultural Societies; for the extent to which the institution can be rendered efficient, depends upon the support which it meets with, and the funds which are raised for carrying it into execution. It is sincerely to be wished, should the honourable Board approve hereof, that Parliament would, upon an application, see the expediency of granting the funds which may be necessary for an undertaking so highly conducive to the public welfare. Were one great National Institution established, the Agricultural Societies in the different districts would the more readily lend the assistance requisite to try such experiments as the peculiarity of their situations may afford towards completing the design.

THOMAS BATES.

Halton Castle, Northumberland,
December 19th, 1807.

PART IV.

Mr. Bates as an Improving Agriculturist and his intimacy with the Leading Men of his Time—Mr. Curwen's Correspondence—Mr. Mackenzie's Correspondence and Death—Mr. Trevelyan's Letters on an Experimental Farm and Agricultural Chemistry.

When Mr. Bates entered upon Halton Castle farms, he cultivated turnips to a very great extent, acting upon his principle that the green crops were the foundation of all good agriculture. He introduced all the newest and best varieties; but he had great difficulty in obtaining a proper farm steward to carry out his views, until my father, Robert Bell, on the recommendation of Mr. Hopper, of Black Hedley, took charge of the farms. My father enjoyed the confidence and friendship of Mr. Bates until his death in 1848, having removed to Kirklevington in 1822.

Mr. Bates was in constant communication with the leading agriculturists in all parts of the kingdom, and he attended most of the great gatherings; and for several years he kept up a close intimacy with Mr. Curwen, and was visited, at Halton Castle, by agriculturists from all parts of the kingdom to look over both the cattle and tillage crops. Lord Strathmore was a frequent visitor, he being one of the leading breeders and providing bulls for the use of his tenantry, and thus enabling them to keep stock of the best quality.*

In October, 1807, Mr. Bates had an auction of his

---

* Many excellent short-horn cattle are now to be met with in the Vale of the Derwent. Mr. G. H. Ramsay has many cows with the original Duchess red and white marks; but not having kept pedigrees, he can only trace them to the use of Lord Strathmore's bulls. *The Author.*

surplus stock, numbering sixty: but that mode of sale being then little known, the result did not answer his expectations. The dairy was carefully attended to by Mr. Bates. Large quantities of excellent cheese were made at Halton.† A market cart also went twice a week to Newcastle with the butter, which sold much beyond the ordinary market price. Mr. Bates made various experiments to test, not only the quantity, but the quality of the milk produced by the different kinds of cattle and food. Mr. Bailey, in his "Survey of Durham," refers to them, but Mr. Bates was not satisfied with the accuracy of Mr. Bailey's statements, and wished them corrected or suppressed, and Mr. Bailey complied with his request in the supplement to his book, page 409.

Sir John Sinclair, whose friendship Mr. Bates enjoyed in Edinburgh, wrote him on the subject of dairy produce, as follows:—

"It is mentioned in the Durham report that you would favour the Board of Agriculture with the result of some experiments for ascertaining the proportion between milk and butter. It would be obliging in you, therefore, to send that information to me as soon as may be convenient, under the cover of the Earl of Caithness, Edinburgh. I should be glad also, that you would make comparative trials of churning the *whole milk*, both sweet and sour, against churning the *cream alone*. Also on the different value of the buttermilk of each, for feeding pigs. Any information from you will always be acceptable to the Board of Agriculture."

The noblemen and gentlemen who took such an active and patriotic interest in the improvement of agriculture

---

† I believe that no cheese whatever is now made in the County of Northumberland. The ordinary cheese used by the labouring classes is, it appears, nearly all American. *The Author.*

and stock at the beginning of the century had a royal patron and fellow labourer in His Majesty King George III. ("Farmer George," as he was so often called). His Majesty probably devoted most attention to sheep and wool. His Spanish Merinos and the crosses obtained great perfection, both in flesh and wool. His Majesty, when at Weymouth, visited all the best farmers' flocks, and very frequently discussed the subject with the owners. Many anecdotes are still recollected of His Majesty in such visits. The large importations of Merino wool from Australia destroyed the commercial value of the Merinos in England, although, I believe, some flocks still exist in the eastern counties. Mr. Western at one time had an excellent flock. His Majesty obtained many excellent short-horns from the north. Portraits of His Majesty's cattle were published in 1801, viz., a bull, cow, and ox, and also a portrait of the Ketton or wonderful ox, which were then all called Holderness cattle.* Windsor (698), was hired by His Majesty, and hence his name. He is mentioned afterwards.

Mr. Curwen, M.P., of Workington Hall, Cumberland, was one of the most patriotic and successful improvers of cultivation and stock in the north. He was a frequent visitor of Mr. Bates's, at Halton Castle, and they were in constant communication on agricultural subjects. Mr. Curwen carried out many experiments according to the ideas of Mr. Bates, as given in his letter to the Board of Agriculture.

* See Garrard's prints of Improved British Cattle, published at the Agricultural Museum, George Street, Hanover Square, London, 1801.

Mr. Curwen made a Report of the results of his fattening experiments, for which the Board of Agriculture awarded him a £50 prize. His "experimental cattle," consisted of a couple of short-horns, Herefords, Glamorgans, Galloways, and long-horns, and a solitary Sussex. The greatest profit was £8 10s. 1d. on short-horn, No. 2, which increased in weight from 90 stones to 115 stones; and the second best was £6 16s. 5d. on a Hereford, which began at 61st. 7lbs., and made 28st. 7lbs. In the case of the former, the food, in which 6st. 6lbs. of oilcake was the only artificial stimulant, cost £7 17s. 7d., and in the latter £7 19s. 11d.; and each of them was purchased at 4s. and sold at 6s. per stone.

Mr. Bates attended the annual gatherings at Workington, which was then the Holkam or Woburn of the north. At these meetings the farms were inspected, new implements exhibited, and new manures discussed. Professor Coventry's writings were then the text for chemistry.* The works of Professor and Bishop Watson of Cambridge, were also much studied. Mr. Curwen paid great attention to the dairy qualities of his stock. His published work had special reference to the supply of milk for the poor,† and in his preface he said, "A more rapid revolution of public opinion had already taken place than had been observed in Great Britain within these last few years, in the universal estimation and importance

* Discourses explanatory of the object and plan of the Course of Lectures on Agricultural and Rural Economy, by Andrew Coventry, 1808.

† Hints on the manner of Feeding Stock, and the Condition of the Poor, by J. C. Curwen, M.P., of Workington Hall, Cumberland.

attached to agriculture." From a mass of correspondence, I have selected the following letters to show what the opinions and exertions of patriotic agriculturists were at that period. They were written to Mr. Bates when in Edinburgh. He had then bought the Kirklevington estate. Mr. Bates had a special method of preserving turnips.

"Workington Hall, January 5th, 1811.

"My dear Bates,—What with indisposition, and what with the effects of vexatious and untoward circumstances, I have been very ill and uncomfortable for the last three months. A most serious loss was sustained from fire on my new winning. It threatened its total destruction; but providentially not a life was lost, and fortunately not a tenth part of the damage I had reason to expect. Since that, the bank of Messrs. Bowes, Hodgson, and Falcon has stopped payment. This occasioned a most severe run upon Messrs. Woods and Co.'s bank, and a great stagnation of all trade. Though my individual loss is trifling, that of the public is considerable, and it will lock up £40,000 for a length of time, which will be greatly felt by all persons in trade. This will account for my suffering your letter to remain so long unacknowledged. I hope you continue pleased with your purchase. I think what you propose can scarce fail of answering. Land will, I consider, advance yet greatly in value. There is an estate to be sold at Harnington, of which I have the royalty—80 acres with a good house. They ask £10,000. I shall not be surprised should they get £8,000. I have been compelled to take a hand in helping other people to breed short-horns; I have got a bull from Mr. Mason; my situation precludes it; a calf at three months old costs me five guineas. If I can encourage other persons, I shall be satisfied. I had 95 gallons, 3 quarts, wine measure, churned yesterday. It gave 30lbs., of 16oz., or 17oz., on three gallons. This, considering the severity of the weather, is not doing ill. I have been very actively employed on my farm. I am ambitious of showing it next year in a state not unworthy of the notice of my Scotch and South Country friends. My turnips keep well, thanks to you. Several farmers have imitated the example set them. The being able to have them unfrozen is a great matter for dairy cows. My wheat is very good, but does not yield above 26 Winchesters per acre. It weighs 61lbs. I hope you will very shortly receive the report. It will require great allowances. My patience has been exhausted by the printer: I wish yours may not by the author. I do not know how you will like what I have said respecting stock. I have spoken my real sentiments, pretending to no knowledge on this subject. Mr. Maynard, jun., has surprised me. I hope the result may justify him. When you see any of my East Lothian friends, pray remember me to them. You will probably know how the Scotch review of the reports goes on. I see Sir John is supporting the Regent, though he was made Right Honourable by Mr. Percival. Gratitude in politics would be a monster, and, something like honesty, out of place. I fear I must go to town soon. Mrs. Curwen is much obliged for your present of a book, and very much pleased with it, and I know of no better judge. I shall be very happy to hear from you. If you write, do so within ten days; otherwise I shall be on my road to town."

"Workington Hall, Jan. 12th, 1811.

"Thanks for your letter. I rejoice to hear from all quarters so favourable an account of your purchase. I hope it will do more than realise all your expectations.

"I could almost scold you for supposing me so much prejudiced that I should not be disposed to make a fair report. Of your cross, I have no experience, I must therefore speak with diffidence. I have, to the best of my belief, stated the question fairly as to the objects you have in view. That the short-horned are an admirable breed of cattle cannot be questioned; that they are at the utmost point of perfection, no prudent man would contend. As milch cows I do not believe they have their equal. With respect to butter, that is another question. That will require more experiments to decide than have yet been made. I can procure 20oz. of butter from 12 wine quarts on particular food; 17 is the least. I did not try the experiment at the height of the clover. Oil-cake makes a great difference. I am disposed to think they eat more than the long-horned. I have not a milch cow which is not fit for the butcher. Here is an advantage I never saw in any other cattle—to milk, work, and lay on fat at the same time. I hope, when you have seen the report, you will be of opinion I have dealt fairly, though I have no hesitation in declaring that I am strongly of opinion your cross will never rival the short-horned. Both truth and fair statement is due to every subject; and, whilst I think one way, I most sincerely wish I may be mistaken, for the sake of my friend. This is all that candour can expect or you would require. A few weeks will give the report; and if I have erred, I will most readily give every publicity to any statement of yours next year. I have always sought truth, and been anxious for the public benefit. Every experiment is entitled to the greatest consideration. The individual risks: the public can alone be gainers. I highly disapprove of any man lending himself to serve any individuals. Though you are not in the County of Durham, your experiment may be of such public utility that it ought to have every fair play. My contention is not as to the kyloe but the long-horned, the worst breed of cattle in the island. Here, then, the matter must rest till you have seen the report, when I shall be much obliged for your candid sentiments. No man more likely to be in error; but, at the same time, I am not so attached to my own opinions as to hold them a moment longer than they appear sanctioned by argument and fair deduction.

"I go to town in a week. Reluctantly I quit my farm. My turnips keep most admirably, and have fed during the frost without being in the least affected, which is important for a dairy.

"P.S.—I expect Mr. Mason's bull on the 17th. It is for the country, not myself."

"Workington Hall, Feb. 11th, 1811.

"The report is at length finished, and shall be sent you either to-morrow or Thursday. I send one for Dr. Coventry, Mr. Graves, Lord Woodslie, Mr. Browne and Rennie, with a second for yourself. I must beg the favour of you to send them to the different parties. I shall be very anxious to know your sentiments, whether I have stated your objects fairly. As a cow-keeper, I am very partial to the short-horns, and when there is plenty of good food and no exposure, I do not think for that object they can be surpassed. Situation makes a great difference. The country appears much pleased with Mr. Mason's bull. I am satisfied we shall agree in opinion that the long-horned cattle are the worst breed extant. Many of my farmers are getting rid of them. I sent Mr. Gibson's man an order for sixteen heifers for my tenants. The printer has just shown me the last proof, so that seven copies will be sent to you by the coach. If you can, spare one for General Dixon. I will send you more if you wish for them. I got to town on Saturday, where I shall be happy to hear from you. Dreadful weather. The wheat has suffered a good deal in various parts. I have not seen Spedding for a length of time. I shall call upon him as I go south. My health has been very indifferent, but I think I am better within this week. I beg you not to spare the report, but tell me your real sentiments. I shall be very happy to correct anything that is erroneous. If I get away from town early in May, my first visit shall be to Galloway, Ayr, and then to Edinburgh,—I hope I may find you there— Returning by the Lothians. Till this moment I have forgot Hepburn; I will therefore send and order ten, which leaves you two still to dispose of. Your method of preserving turnips has answered completely."

Mr. Curwen expected each dairy cow to make £25 a year.

Mr. Bates travelled much, and had a large circle of friends at great distances. In 1809-10-11, he spent the winters in Edinburgh to attend the Professors' Lectures. He saw much of the leading Scotch agriculturists. His note book of the lectures of the Professors of Chemistry and Mineralogy are still preserved. To these visits he, Mr. Bates, owed one of his most pleasing friendships, viz.: with Thos. Mackenzie, Esq., many years M.P. for Ross-shire. The introduction, so fortunate to both parties, was by the following letter, in March, 1812, from Dr. Robert Jameson, the Professor of Mineralogy:—

"I find it will not be in my power to get the length of Halton Castle this summer. I enclose to you a letter from my particular friend Mr. Mackenzie, junr., of Applecross, one of our great Highland proprietors, who is very desirous of making himself acquainted with the husbandry of your part of England. The letter will explain his wishes, and I trust you will be able to grant his request. I have known Mr. Mackenzie for several years, and have found him a man of the strictest honour, of most amiable manners, and deeply versed in all the sciences in which you feel an interest. In short, my dear Sir, should your domestic arrangements admit of an additional inmate, I am convinced that the only inconvenience you will have to experience will be the regret on parting with a gentleman of Mr. Mackenzie's talents and worth. I hope you have not renounced your mineralogical pursuits. They will always prove an agreeable relaxation, and may, to you, afford means of improvement. My friend, Mr. Mackenzie, is well skilled in mineralogy, and expect that together you will be able to give me a correct idea of the structure of the North of England."

Mr. Mackenzie, on the 31st March, 1812, wrote to Mr. Bates as follows:—

"Our excellent friend, Professor Jameson, having just communicated to me your letter of the 28th ult., I make no apology for addressing you myself, to assure you how much I feel gratified and obliged by its contents. My view in wishing to pass some time this summer in Northumberland was solely the acquisition of so much practical knowledge in husbandry (which I have sought in books in vain), as might enable me to carry on judiciously the improvement of a large highland property (at present nearly in a state of nature), which my father leaves entirely to my management. From having seen you occasionally at Dr. Coventry's class, I had the highest character of your talents from my friends, Mr. Jameson, Mr. Erskine, &c. I immediately thought of you as a person not only eminently qualified to advance my agricultural knowledge, but from whose general information I should derive both pleasure and instruction. Mr. Jameson obligingly offered to communicate my views to you, and I consider his having introduced me to your acquaintance as far from the least of the many favours I have received from him. I think I shall be able to arrange my affairs here, so as to be at liberty about the middle of next month, when I shall do myself the pleasure of visiting you; and should it not suit you to gratify my original wish of residing some months under your roof, I shall at least have the advantage of your advice as to the best means of attaining my object. I shall endeavour to meet the Mr. Campbell you mention, and when I know with certainty how soon I can leave Edinburgh, shall do myself the pleasure of writing you."

Mr. Mackenzie resided a considerable time at Halton Castle, and frequently visited there. He kept up a constant correspondence with Mr. Bates. Mr. Mackenzie was familiar with all the best cultivated districts in England and Scotland; and when at Halton Castle he made himself master of all the details of the management of the farms and stock then under the charge of my late father, Robert Bell.

Mr. Mackenzie wished to introduce the cross of Mr. Bates's bulls with the highlanders. Mr. Mackenzie and the other proprietors obtained bulls from Halton Castle. Mr. Mackenzie entertained so high an opinion of the management of affairs at Halton, that he wished his agent to be instructed there, and he was resident there for sometime. The following letter from Mr. Mackenzie to Mr. Bates explains the object:—

"Applecross House, 25th April, 1817.

"My dear Sir,—I was this day favoured with your letter of 17th ult., and return my sincerest thanks for your kindness in allowing me to send the young man, who is to be my bailiff, to Halton, and for the valuable information which it contains of markets, &c. I write the young man to-night, and enclose this to serve as his introduction, as you are kind enough to promise to direct his attention to such points as may be principally useful to him in my employment.

"I shall briefly mention my object in sending him to you. For the first year he will have little to do with me, as the outgoing tenants have the way-going crops, but afterwards he will have the whole charge of the culture and improvements of the farm under my direction. The buying or selling of stock I generally contrive to manage myself. Were I determined to keep only highland cattle I could myself instruct him in the detail of their management, but as I have a great inclination, when my farm is in order, to try the mixed breeds, I wish him to be exactly acquainted with the minuteness of their management, of which, you know, I am ignorant, and to gain some knowledge of stock; but I am above all desirous that he should see and imitate the regularity of your system, not only in work, but in the habits of your workmen, and form himself on Robert Bell's model as nearly as he can. You know me so well as to be aware how little I could tolerate an immoral servant, or one who had any high notions. The former I am assured he is not, the latter I think he can in no way be better guarded against, than by seeing your system under Robert Bell. As he has not been accustomed to have the command of workpeople, he might run the risk of mistaking presumptuous domineering for proper authority. I think you will see my object; perhaps you will also permit him to see how Robert keeps his books. I am just setting out to attend a County meeting."

In 1822, when the agricultural interest was in such a depressed state, Mr. Bates vistited London, and endeavoured to arouse the landed proprietors to obtain some relief from the Legislature. Mr. Mackenzie's last letter to Mr. Bates, dated August 3rd, 1822, from Applecross, paid a great tribute to the exertions of Mr. Bates:—

"Your letter from London after going to Brighton found me here. I cannot tell you how much I regret your not having arrived in town before I left, as few things would have afforded me such sincere pleasure as to shake hands with you once more, and perhaps, from my acquaintance with parliamentary business, I could have saved you some trouble. *If all who are interested in agriculture had had a small portion of your zeal and activity, matters would be very different from what they are; but I dare say you have seen enough to satisfy you of the apathy of great men, in regard even to what most dearly concerns their own interest.*

"The plan which you sketched out for my journey northwards would have been quite delightful, but I fear would not have suited my state of health. I came from London to Leith by sea by my physician's advice, and thence here by very easy stages, where I am living in perfect retirement. I have now no complaint whatever, except weakness from the medical treatment which I suffered. If I can regain my strength during the fine weather, I shall, I trust, be able to resume my duties in winter, and shall endeavour so to arrange as to visit you; but if I do not get strength before the cold weather sets in, I have little hope of being able to stand it, and must look forward to our meeting in another and better world; whichever may be the case I am content, being satisfied that whatever the Almighty decrees is not only the wisest, but in every respect the most beneficial fate for his creatures.

"When you reach Ridley Hall I shall expect a few lines from you, as I shall be anxious to hear that you have not suffered from the fatigue which you have undergone. Our crops are beautiful, and promise to be early, except hay, which was very bad; no demand for cattle at any price. Wool very low. *I know not what is to become of tenants and landlords unless times mend soon.*"

Mr. Mackenzie soon after returned and died at Brighton.

Sir John Sinclair, Bart., quotes the opinion of his friend and correspondent, Walter Trevelyan, Esq., of Netherwitton, near Morpeth, "that the farmers of Scotland are too apt to overlook and underrate their own breeds of cattle."*

* An account of the system of husbandry in Scotland. Edinburgh, 1812.

Mr. Trevelyan was an intimate friend and correspondent of Mr. Bates. I insert two letters of Mr. Trevelyan, of Netherwitton, to Mr. Bates, in Edinburgh, to show the attention then paid to agricultural chemistry:—

"17th June, 1810.

"Dear Sir,—Appearances of my neglect towards you are certainly much against me. I wrote a full answer to yours of April last, but had not, upon re-perusing it, the resolution to send it. The letter being hardly suitable in regard to subjects to be presented to the Chancellor of the Exchequer, in the House of Commons, my letter being fitter for an after-dinner reading than before it, which I will preserve for your reading here. If you will do me that *particular* favor on your return, on your way home, this house being on the shortest and best road from Edinburgh, but give me a previous line lest I might be out of the way. This day week I dined at Matfen—Ladies Collingwood and Blackett being present—where I met your brother John. He is not an agriculturist nor a horticulturist, although he speaks of having a garden of 20 acres, for the Newcastle markets, on some ground on the North of the Tyne. The drought has been severe and continuous, so I have been looking to my dung pits, that the manure, for the drill turnips, may be as moist as can adhere to the forks. Last week I went from Newcastle, Tyneside, to Haydon Bridge. I do not believe that in any part of England, this county excepted, there is so great an extent of horticultural rather than agricultural tillage to be seen. We certainly take the lead nearly out of eye reach. An *experimental farm*, once my hobby, would place us amongst the celestials. you and I may pull together—and if sadly jeered at by the over wise, we shall not be altogether derided. I could say much more, the subject being fathomless, and shall add only my sense of your attention towards me, and that in all your agricultural views you may command the assistance of

"Sincerely yours,
"W. TREVELYAN.

"P.S.—There is a Dr. Miller, of Fountain's Bridge, Edinburgh, chemical professor, who owes me much in his professional line. Pray call upon him in my name."

And in January, 1811, Mr. Trevelyan writes as follows:—

"Many thanks for your information, on the benefit to be derived from liming clover after the first cutting, which I shall particularly attend to this summer. What is your opinion of liming clover lay just previous to the ploughing for a spring crop? I should think it right, as the alkaline qualities greatly assist the fermentation of the roots, &c., by decomposition, being as it were an artifical hot bed, the steam being a greater promoter of vegetation than is generally understood, or rather not understood at all: the steam, being impregnated with the volatile oils of manures through which it passes, is fitter for entering the innumerable pores of the plants. To enter at large on this subject, as to the ascent of steam and vapour, or the principles of vegetation, would ramify into such various branches as to require sheets of paper. You remark that 'lime operates many different ways, and that the proper time of applying it has not been duly attended to.' I have applied it to great advantage upon coarse grass, which cattle would not touch, particularly where the cows are usually milked, which is owing to the super-abundance of the saline and nitre acids, which being neutralized by lime, corrects it, or reduces it to its natural sweetness, genial to the palate or taste, and, therefore, I use quantities of lime in my composts, which are made of the overflowings of the foldyard; much loss is often sustained by allowing the rain to wash away the saline particles of the manure. I request you 'to enquire, what are the best tests to detect the component parts of limestone which are injurious to vegetation.' We well know the difference beds or strata of limestone do not possess alike the genial properties of vegetation, for which cause, I believe, so many people differ in opinion as to the good or useless, perhaps negative, effects of lime. Shells surely may be said to be the purest of calcareous alkalines. Some say magnesia is injurious, which I do not rightly comprehend, as it certaintly is a corrector of the sereline and aluminous acids. What a field, my good Sir, does this open for the improvement of morasses and wastes. I again request, as you are at the fountain head of chemical information to make these enquiries which are of so much consequence to farmers. Are you of opinion that the new grass of new laid or sown uplands is a promoter of the rot in sheep? Every stock farmer asserts it; I cannot believe it, where the bottom is dry. This is a great argument with some against water meadows, but I will tell you a fact, that my brother's Park, at Nettlecombe, in Somersetshire, is watered all the year through, and sheep and deer are perfectly sound; yet I observe the meadows where the very same water passes through certainly rot. You have judged perfectly right in purchasing land. The mind of no *thinking* tenant upon lease can be at ease; but I truly lament that you did not purchase in this county. My mind is so full of agricultural ideas, that unless you take me in your return from Edinburgh, they will ever be in a fermenting state. Be assured I will take an early opportunity of visiting you in Yorkshire, and flatter myself that you will not find my company troublesome. I therefore again request your promise to take me in your way for the day only, and I truly lament that we were not earlier acquainted, as I knew not the value of that mind which you possess. The company of farmers in regard to tillage, combined with chemical effects, are as empty as full blown bladders, and they dispise reading the experiments."

PART V.

Mr. Bates wanted the Security which Belongs to Owners of Land—Purchased Kirklevington—Lord Althorp and Mr. Bates and the Duchess Cattle—Mr. John Grey and Star, North Star, and Sparkles—Lord Althorp's Correspondence on Short-horns, and the Politics of his day—Mr. Champion bought Stock of Messrs. Colling, and Lord Althorp on his Herd—Lord Althorp's and Mr. Grey's Herds—Lord Althorp's view to the "Main Chance"—Lord Althorp and Mr. Bates become Less Friendly, and the Reason Why—Lord Althorp and Mr. Bates were Actuated by Different Motives; the Result being the Former was not a Successful Breeder.

Mr. Bates was not satisfied with his position as tenant of the land he occupied. He had no expectation of being able to increase his estate at Park End, and he considered that it was not capable of much further improvement. His farm at Halton and other pursuits left him no time to attend to Park End. But he wished to purchase a large estate for retirement at the expiration of his lease of Halton. Mr. Bates said, the Park End estate was purchased in 1663 for £223 5s., and sold in 1703 for £1000; and that seventy years afterwards it was offered for sale, and it was fourteen years in the market before it was sold to my father for £4,000. The only other offer was £2,500; and when I commenced farming it in 1796, Mr. Thos. Ridley, who had made offers, said he would not give more for it. The first three years I occupied it, it did not pay the expenses of cultivation and management, but it improved so much that Mr. Thomas Ridley offered £14,000 for it. It was ten miles from any market town, and possessed no local advantages, and was depreciated in value by a new claim for tithes which his father would not litigate. Mr. Bates, who had been presented with the estates by his father in 1806, sold it to Mr. Thos. Ridley in 1812. Mr. Ridley, who had farmed it himself, said it was cheaper than if he had got it in its unimproved state for £2,500.

Mr. Bates, who had largely increased his fortune in 1811, purchased the estate of Kirklevington, near Yarm. It is pleasantly situated on rising ground in the Vale of Cleveland, and mostly on the new red sandstone formation. It contains some excellent grass land. Kirklevington had been the seat of the Percys, in Cleveland, and afterwards belonged to the Strathmore family, and was many years occupied by the Maynard family, so well known in short-horn history. In 1818, Mr. Bates purchased the Ridley Hall estate in South Tynedale. It had an excellent house, but only about 300 acres of land; and on leaving Halton in 1821, he went to reside at Ridley Hall and continued there until 1830. Before purchasing Kirklevington he had bought the Hall Garth estate near Durham, but the title was defective in respect of tythes. The Vicar of the parish made claim for the tythe on Ridley Hall, which Mr. Bates thought inequitable, the Vicar had as trustee sold the estate to him as not liable to tythe, and after some disputes and litigation Mr. Bates sold the estate and then went to live constantly at Kirklevington.

I cannot now ascertain when Lord Althorpe and Mr. Bates first met. They certainly met at Mr. R. Colling's sale in 1818, and the kindred spirit in short-horns soon made them friends. His lordship is too well known to all agriculturists to need any mention here. In the political world he is best known as the Chancellor of the Exchequer in Lord Grey's ministry—1830 to 1833. His short-horn herd at Wiseton had long been established, and he was also a successful exhibitor of fat cattle at the Smithfield Club Shows in London for Christmas beef. Mr. Bates freey delivered to his lordship his opinion of the Duchess cattle, and impressed upon his lordship the importance of using that blood which then existed in Northumberland through Duke (226). His lordship had visited Mr. Bates in 1818, or early in 1819, and Mr. Bates purchased for him several short-horns of the Duchess blood. Writing in 1819, his lordship says:—" If the Ben (70) blood is as bad as you seem to think, it is very difficult to avoid it, for, if I remember rightly, it comes into Windsor's (698) blood, and therefore into Star's. Pray what is the disease, and what cattle of this blood are subject to it? I know that

Mason's St. John (572) sort have weak constitutions, but I do not know of any particular disease. Though I am very eager to see Spot and Sparkles, yet I shall not be over at Wiseton till the end of September. I think it will be best for me to accept your kind offer to keep them for me until cool weather. I think you have made an excellent bargain for me, for the prices are very iittle more than butcher's prices." Spot was bred by Mr. Compton, the purchaser of Duke (226), and was by Duke, dam Premium cow (page 411 Herd Book, part I.), and was the dam of Ferdinand (258). Sparkles, or Sparkler, I am unable to trace, except that she was bought of John Grey, Esq., of Millfield Hill, and afterwards of Dilston and Commissioner of the Greenwich Hospital estates.* Star belonged to Mr. Grey, and was I suppose a son of North Star (458). His fate appears in the following letter of Mr. Grey to Mr. Bates:—

"Milfield Hill, Sept. 1st, 1819.

"Dear Sir,—I intended to have written to you in a post or two to acquaint you of the time when Sparkles was served, and to arrange about her going home, when I had the pleasure of receiving yours from Yorkshire informing me of the chance of Lord Althorp doing me the honour to call on his return from the Highlands. Should his lordship find time to do so, I shall have much pleasure in showing him anything that this country can afford that he may deem worthy of his attention. I think I might venture to say, that I could show him on my farm 100 acres of as good turnips as he would find in any man's possession from the Tweed to the Thames. But it is a subject of no ordinary regret with me to state that I cannot show him Star, as that valuable animal died last week. I found on my return from your neighbourhood that he had been very ill during

* *Vide* the memoir of John Grey, of Dilston, by his daughter, J. E. Butler. Edinburgh, 1869.

my absence, and that in spite of every attempt to relieve him, he continued growing worse from day to day and entirely forsook his food, so that he was supported for some time by gruels, &c. He never had any difficulty of breathing, but seemed to suffer much pain internally, and when he died at last quite exhausted found the seat of his complaint to be in the liver, which was covered in parts with small hard knots of a whitish colour, which also was fastened to his side. His throat, heart, and lungs were quite clear of any appearance of ulcers or disorder of any kind. The complaint cannot have been of very recent origin, but it is probable that the extreme heat of the weather, by which he seemed to be excessively distressed, had brought it sooner to a crisis. I wish he had lived one year longer that some more varieties might have been produced from so pure a stock,—but all wishes and regrets on that head are now equally unavailing. Sparkles was served by Star on the 25th June, and again on the 16th July; since which time she has never come in season, from which I presume there is no danger of her. I was going to propose to you, when you wish them to go home, to allow your man, who will no doubt be a careful driver, to bring along with him three cows that I have at Mr. Donkin's, for doing which, I would of course pay his day's wages and expenses; or if you prefer it, I could send a man from here with Spot and Sparkles who would bring mine home. If you determine on the former plan, you can send them off at your own convenience; but if the latter, be pleased to let me know. Our turnip crops were beginning to feel the severe drought, but have now abundance of rain, and are generally the best I ever remember to have seen. Everything is much refreshed by the change of weather; although in this stage of the harvest, many would have willingly dispensed with rain for a week or two. We are later this year than usual in proportion to other parts, little has been stacked yet, and stands so deplorably wet in the fields. Oats are but a light crop everywhere. Spring wheats generally so, but well filled. Winter wheats will not yield to the bulk—the heaviest crops are the worst grain. Barley is good. I have heard of no ewes being sold, but two or three parcels to go to the neighbourhood of London, at from 40s. to 44s., about the same as last year,—they are commonly thin in condition. There is no market for dinmonts yet, but as turnips are good I trust the prices may keep up. I send for Spot to-day. Mrs. Grey joins in kind regards with, dear Sir,

"Yours faithfully,

"JOHN GREY."

Lord Althorp wrote to Mr. Bates the following letters:—

"My dear Sir,—In consequence of my absence from home, I did not write to you according to your desire at Halton Castle. I shall now send a duplicate of this letter to you there, at least of so much of it as is of any importance that you should receive immediately. I should very much like to have Sparkles, and the 30 guineas cow of Mr. Hutchinson's home as soon as I can, and as Mr. Blackett has not called here yet on his return, I hope I shall be in time to have Sparkles come to Kirklevington to meet him. I am very full stocked, so that I think it better to decline Mr. Huchinson's old cow, and Mr. Place's heifer, for the 30 guineas cow will give me a cross of all the blood, at a moderate expense. I am very sorry for Spot's accident, but I hope, by rest, she will soon recover; as to your standing to the loss, we will talk about that if it should occur. I must, however, ask you to let her stay at Halton till after her calving, as moving her before will now be impossible, and perhaps you can have her put to Duke.

"Parliament is to meet, as you know, on the 23rd of next month, which will prevent my Northern expedition before Christmas, but if the recess is long enough I will come afterwards.

"Yours most sincerely,
"Wiseton, October 30th, 1819."   "ALTHORP.

"My dear Sir,—I am on my way up to town, and as I must on getting there drive short-horns out of my head, to replace them by politics—certainly a more disagreeable subject, and perhaps in the present state of things a less useful one—I write you a parting letter to say, that I suppose Rosette (page 475, Herd Book, vol. I.), will have arrived when you receive this, and she may be put to the bull, the first time that she is ready. I think I told you that I wished you to bleed her when she arrives. Sparkles has had a bull calf; I have not seen him, but I hear he promises well. I cannot, however, be very sanguine about him, for I think her coarse shoulders must come out in a bull, but I shall not have him cut till I see him in the end of July. By the bye, I do not know whether you would like to see our Cattle Show, at Doncaster, which will take place about that period; if you should, I need not say, I shall be most happy to see you at Wiseton, and if you will say so, I will let you know the day in time.

"Yours most truly,
"Dunstable, April 26th, 1820." "ALTHORP.

Lord Althorp hired His Grace (311) of Mr. Bates, and wrote as follows:—

"My dear Sir,—You may say what you please, but I will say that I am very much obliged to you for the inclination you show to accede to my proposal about His Grace. I am thus situated; I have a young bull that I expect will be good enough to put to all the common Regent heifers next year, but I hardly expect that he will be good enough to put to a bull breeder, and therefore if Nonpariel should have a heifer calf, I shall not exactly know what to put her to, because my young bull will not be worthy, and Regent is not correct enough in his shape to be put to his daughter, but even in this case she will be fit to put to the bull before the 1st August. I would therefore propose to take His Grace for a twelvemonth, or to the end of the year 1822, from the 1st August, 1821, if Nonpariel has a heifer calf; if she has a bull calf, then I will take him for the year 1822. I do not consider myself bound to allow Champion the use of a bull I hire, as this is a very unusual thing to be allowed, and I should also rather prefer that you did not allow me to take in cows to His Grace. This being the nature of my proposal, I shall be obliged to you to let me know what you will ask for the hire of him for the periods I have mentioned.

"Yours most truly,
"St. James's Place, May 1st, 1820." "ALTHORP.

"My dear Sir,—I write you a line to say that I have bought Premium (page 441, Herd Book), and I have desired Mr. Grey to send her to Kirklevington. She has only calved about a month, so she will be very fit to travel home with Palm-flower (page 429). When she arrives, please write to Hall, at Wiseton, and he shall send for the two. I only gave £31 15s. for her, so she is very cheap.

"Yours most sincerely,

"St. James's Place, May 15th, 1820." "ALTHORP.

"My dear Sir,—I arrived here last night and found His Grace arrived safe, but not quite well as he coughs a little. I have therefore ordered him to be blooded pretty copiously and to have a dose of physic. If this does not remove the cough, I shall apply the blister which succeeded so admirably with Spot.

"I like him very much indeed, and a great deal better than I expected; indeed, I can only find one point to criticise, and that is that the arm of his foreleg is coarse and too muscular. This, however, is of no great consequence, and I have no doubt from the perfection of his shape and quality in every other respect, that he will do me the greatest service; I use the word perfection because I think it most appropriate to the merits of His Grace. I wish you joy of the birth of Earl Percy. Duke has scarcely more of his own blood than the young Earl has, and therefore if he turns out well he will be very valuable indeed.

"I am afraid your plan of a ministry composed of men of all parties will not succeed. The members of the administration would have no confidence in one another, and the public, who would attribute their union to interested motives, would have no confidence in them either. This has always been the case in all coalitions which have ever taken place, and I think probably always will. The first thing to give energy to a government is that its members should have confidence in the honour and integrity of one another; the second is, that they should agree in the main principles on which their political opinions are founded; and without these two qualifications no government can be good for anything. The ministry which you propose would probably not possess the first of them, though perhaps it ought, and certainly would not possess the second. For these reasons I cannot agree with you in wishing the experiment to be made.

"Believe me, my dear Sir,

"Yours most sincerely,

"Wiseton, Nov. 16th, 1820." "ALTHORP.

Mr. Champion was a well known breeder of cattle, and purchased many for Messrs. Colling; and Lord Althorp, writing to Mr. Bates, refers to his herd and other matters in the following letters:—

"My dear Sir,—I am happy to tell you that His Grace's cough is nearly well. He has never appeared ill at all with it, and has shown such a constitution as I never saw in any beast before. No physic will have any effect upon him; I was able only to relax his body a little by three pints of castor oil, for salts, the medicine I usually give, would have no effect upon him. Upon consideration, I thought it best not to apply a blister, for I was afraid it might irritate him, and I am always very cautious about a bull's temper. I like him better every time I look at him, and I think I should say he is the finest bull I ever saw in my life, not excepting Duke himself. Champion has seen him and admires him very much, but has, as I expected, refused your offer; you will not be surprised at this, when I tell you that Empress is now in calf to the young brown bull with a black face, which he showed at Doncaster. He is determined to have a new sort of his own; I told him that they might probably be a better sort of cattle than the improved short-horns, but that most undoubtedly they would not be improved short-horns, but something else. I have not yet seen Simpson, as he is out of the country. I have not seen Champion's young bull, who is matched against Sirloin, but from Champion's manner when he saw Sirloin, I think the match is won already. I hope Earl Percy is well and thriving.

"Believe me, yours most truly,

"Wiseton, Dec. 20, 1820." "ALTHORP.

"My dear Sir,—I have no wish to use His Grace to any more cows, as I have so many in calf to him that I have hazarded enough upon one bull, therefore he will be ready to go whenever and wherever you please. The only cow I should put to him by any chance would be the Lancaster heifer, who has had so good a calf by him; and this, of course, would depend upon how the calf looks when I get back to Wiseton; and this is of no consequence, as I think Regent likely to get as good an animal out of her. I cannot quite give way about the Lame bull. Charles Colling writes me word that he was got by a bull of Brown of Aldborough's, so that on his sire's side at least he was well bred. Then nothing can show better constitution than Peeress, Cecil, and Ketton. Spot is a fine cow, and Premium, her dam, a much finer cow, and these are Lame bull all over. Premium is got by May Duke (Lame bull), dam by Cupid, (Lame bull), grandam by Favourite, out of Johanna, who was got by the Lame bull himself, and I think you told me that nothing could be better than Johanna. Cecil's horns are quite waxy, and so will those of the heifer out of Harrison's cow by him be, and so are Premium's; and, as to the shoulders, in all these three they are the most remarkable points about them; and I should prefer Cecil to any bull I know for a cow who had coarse shoulders. When I told you that the calf out of Rosette was amiss, I did not mean to describe him as actually diseased. He had a scouring upon him for some time, and as this is the disorder most prevalent among the short-horned breed of cattle, I am determined never to use a bull whom I can ascertain to have been affected in this way; that is to say, as long as I can find one that has never been so. This is my only objection at present to Radical. The public admire him very much.

"Believe me, my dear Sir,

"Yours most sincerely,

"Leamington, Sep. 16, 1821. "ALTHORP.

"P.S.—Daisy, the daughter of Sally, belongs to Mr. Simpson and not to Mr. Smith. Sally herself came to Smith's share, and I saw a very promising bull calf out of her by Ketton (Lame bull again), at Dishley, in the summer."

"My dear Sir,—I have had one calf dropped to His Grace since I saw you. It is a heifer out of Rosabella, who was got by Lancaster, out of Rosette. And although it came a month before its time, and although Rosabella is far from being a very strong constitutioned cow, the calf appears uncommonly well, and promises *to be very clever*. I do not know how his having got a heifer calf out of this cow agrees with your system. I am afraid you must either give up your system in this case or His Grace's constitution, and I am sure you need not give up the latter.

"Believe me, my dear Sir,

"Yours most sincerely,

"Wiseton, Sept. 26th, 1821." "ALTHORP.

The system here referred to is, that if a cow is only once served she will produce a heifer calf, but if twice a bull calf. Lord Althorp had paid great attention to such matters, and he believed that there was a popular error as to the time of gestation of cows, and, at Oxford, in 1839, read a paper before the Royal Agricultural Society on the subject.

Mr. Grey had established a good herd, and followed the advice of Mr. Bates in obtaining the Duchess blood. He however appears for one animal only, in the first volume of the Herd Book, viz., Fitzduke (259), by Duke, dam by Phenomenon, grandam Red Rose. Mr. Bates applied to Mr. Grey to obtain him a heifer for a friend who was anxious to exhibit for a prize. Mr. Grey sent for reply:—

"Milfield Hill, February, 28th, 1822.

"My dear Sir,—I was duly favoured with your letter of the 22nd, but cannot as yet give you any information about such a heifer as you are in want of. She should be something superior for such an exhibition. I having been confined entirely to the house for the last ten days by a severe sprain in my leg, have had no opportunity of making any enquiry on the subject. I expect to see Mr. W. Jobson to-day, to whom I will communicate your message, as he may perhaps know of something likely. I believe Hunt, of Thornington, Curry, of Brandon, and some others, have heifers of an age that would suit to show for the sweepstake at the Tankerville Arms, in the first week in April, but they will not probably be visible until that time. I wish you would give us the pleasure of seeing you at that time, and choose for yourself. At any rate, if I hear of anything likely, you may depend on my letting you know. I fear these times will operate against the general improvement of the breeds. Money is scarce with the generality of land occupiers, which will have the effect of throwing a monopoly of the good things into the hands of yourself and such as can afford to go on under any circumstances. I have a bull by Duke that I have used one season, and intended to continue one or two, but am not at all sanguine about breeding, having lost 16 out of 38 of my last year's calves by the quarter-ill during the present winter.

"I wish Ministers had taken your advice and given the bonded grain to feed the destitute manufacturers, or to sell to them at a moderate price, instead of enriching a few speculators at the expense of the agriculturists.

"Most faithfully yours,

"J. GREY."

Lord Althorp had a sale of stock in 1825, and obtained very high prices. Soon after he wrote Mr. Bates the following letter, which shows his lordship had a keen eye for the "main chance":—

"My dear Sir,—I intend to be at Mr. Donkin's. I shall be much obliged to you to let me know where I can see the cows and heifers, descended from Bolingbroke, which you mention. I do not know that I have any peculiar admiration for Bolingbroke's blood, but I do not dislike it as you do, because I cannot think that the sire of Favourite could have been other than a first-rate bull, but I rather want some in-calf cows or heifers, just now to fill up the vacancies occasioned by my sale, and to enable me to have another next year. I can only give moderate prices with this view, and therefore I should be glad to hear where I am likley to find any, in case I should fail in supplying myself at Sandhoe. I am very glad to hear so prosperous an account of your stock, but I beg you not to reserve any bulls for me, as I like two that I have of my own breeding so well, that I do not think I would hire Hubback, or even Favourite, if either of them was offered me. The two I mean are Ivanhoe, by Cecil, out of Brighteyes, of 1817, who is not only a good animal himself, but has proved himself a bull getter."

Mr. Bates and Lord Althorp had a discussion at Doncaster, about the pedigree of the bull Firby, which was much used by his Lordship, and of which Mr. Bates doubted the soundess. Lord Althorp attended Mr. Donkin's sale of short-horns, at Sandhoe, a few miles from Halton Castle, on the 29th August, 1825, and purchased many animals. A coolness afterwards took place between Lord Althorp and Mr. Bates, which I have heard arose from his Lordship having offered my late father, R. Bell, the appointment as his agent at Wiseton, at a salary much beyond what he had from Mr. Bates. But the subject was not mentioned to Mr. Bates, and he first knew of it when my father asked his advice in the matter. My father had continued with Mr. Bates at his original salary, which was very small. Many of my father's contemporaries had obtained appointments at large salaries. Mr. Anthony Todd, who had also, on the recommendation of Mr. Hopper, become farm steward to Mr. John Moore Bates, had gone as head agent to Lord Shrewsbury's estate, and was afterwards with Lord Portman, on his Dorsetshire and Somersetshire estates.

Whatever sympathy there might be between Lord Althorp and Mr. Bates in their devotion to short-horns, there was little in their motives and objects. Mr. Bates never thought of making profit by selling stock, and, in his intercourse with Lord Althorp, had no idea of pecuniary gain. His Lordship afterwards became on very intimiate terms with Mr. John Grey, a more congenial friend in politics. Mr. Bates never forgot the award, by Mr. Grey, of the prize for the best bull in the yard, at York, in 1838, to Lord Althorp's bull Hecatomb, over the Duke of Northumberland, who had got the first prize in his class. This was the only occasion that the Duke was not first when exhibited.

Although Lord Althorp was so devoted and enthusiastic in his love of short-horns, and he had many first-class animals, both in breeding and fat stock, yet as a breeder, I believe, it is now admitted that he was not successful. He produced very few animals considering the magnitude of his herd, and the purchases he made. His stock too were, very many of them, unhealthy or deformed; and he certainly established no tribes or families of stock, and none of the Duchess blood became permanent in his herd, or were propagated from it. His Lordship's case is a proof of the fact that skill and experience is required in breeding. But these his Lordship never obtained. As the defects and failures of his herd have been recently the subject of many articles in print, I do not further refer to them. Mr. Dixon in noticing Mr. Wetherell's Herds (page 173), says:—

"'Great constitution' is Mr. Wetherell's leading tenet, but 'great size' never was, and if he does illustrate it, he goes to Colonel Cradock who gloried in it, and whose 'Magnum Bonum was like the Great Eastern.' He always considers that Earl Spencer (then Lord Althorp) began the bull trade, and made short-horns so to speak fashionable with the landlords. It was the thing to go to Wiseton, more especially about the St. Leger time, and if visitors liked a cow, they bargained to give £50 for the produce. The Earl crossed in-and-in until he sacrificed constitution. They had thin fore-quarters and no breasts; and it was then that Mason, a very clever first-class judge, a hater of 'fool's fat' and open shoulders, and most decided about fore-quarters, and a good neck-vein came to the Earl's aid."

PART VI.

Mr. Bates on the effect of a Fall in Prices after the American War—Lord John Russell a Protectionist—Mr. Bates on Politics—Agricultural Questions—Sir James Graham's, Lord Althorp's, and Sir Thomas Lethbridge's Correspondence on ditto—Free-trade Prices of Corn resulted in Land being put down to Grass—Mr. Bates an Employer of Labour; on the Poor Law Board; and the Turnpike and Highway Question—On the Improvement of, and the Planting of, Wastes and Morasses—Kyloes on improved Waste Lands—Swedish Gentleman's Surprise at our Naked Hills—Mr. Bates on Agricultural Machinery; he advises, in 1838, upon an American Reaping Machine—Mr. Fawcett's Reminiscences of Halton.

Mr. Bates witnessed the distress of the agricultural interest after the first American war. Of this he dreaded a return, and had very strong opinions on the subject of protection to the agricultural interest. It might be an interesting history to record the opinions of statesmen and landowners on this subject. Lord John Russell, when M.P. for Huntingdonshire, had written a pamphlet in favour of protection, and until the subject became a political cry, many of the leading Whig landowners were Protectionists. Mr. Bates witnessed the ruin and misery the low prices of agricultural produce spread all over the North of England. All buildings and improvements were stopped, and landowners with mortgages on their estates, were the first to feel the effects of the reduction of rent, and of increased rates, when they were occupiers themselves. Estates were unsaleable except at prices which left nothing for the owner after discharging the mortgages; and the largest farmers who had leases were soon in the *Gazette*. Many of the leading agriculturists on the Tweed suspended payment, and many who were able emigrated to New South Wales. They were very superior men, and many became eminent agents in various parts of the kingdom. The want of employment was so great that many labourers resorted to their parishes, and thus compelled the farmers and ratepayers to employ them. The system of relieving the poor by payment for labour out of the poor rates was fast becoming as common in the North of England as it was in the Southern counties. Mr. Bates daily received letters with heartrending accounts of distress, and urging him to exertion in getting up petitions for relief and protection, and was in constant correspondence with the leading statesmen on the subject and that of the Poor Laws. He urged Lord Althorp to take a lead in the matter. The Whigs had never been free traders, and C. Fox had once defeated W. Pitt's attempt to obtain a free trade between England and Ireland. Lord Melbourne, as Prime Minister, subsequently declared in the House of Lords that a man must be insane who thought of an absolute free trade in corn.

Mr. Bates also took an active part in relieving agriculture from taxation after the French war. He entered into correspondence on the subject, and urged his views on most of the leading members of Parliament, from whom he received replies generally approving his views. Sir James Graham, Bart., of Netherby, was in constant communication with Mr. Bates on agricultural subjects and improvements. Sir James became a Cabinet Minister in the governments of Lord Grey and Sir Robert Peel: and, on the subject of agricultural distress, Sir James wrote to Mr. Bates in March, 1816, as follows:—

"I received the favour of both your letters, for which I am very much obliged to you, as they are very clear and correspond exactly with my opinions on the subject of agricultural distress, and also on the taxes. I have not been able to obtain a sight of your letter to Mr. Vansittart, but from your letters to me I believe you must have repeated to him nearly what I did in a letter I wrote to him previous to his stating that he should continue the Property Tax at one-half the present rate. I therein stated that if he would confine the tax to 5 per cent. for two years only, and on rent actually paid into the hands of landlords, and give up the tax upon occupiers entirely, and also the tax on agricultural horses, I believed it might pass, although the extent of the landed distress was beyond all belief. But if he proposed to continue the smaller portion of the tax on occupiers, and on agricultural horses, I assured him it would be universally considered as so unjust, unequal and oppressive, that he would lose the tax on property and income, and I have repeatedly pressed this upon the Ministers, but they would persist, and have been defeated, as I foretold them six weeks ago. I perfectly agree with you that the prices of all sorts of produce will in a year increase very much, and perhaps be in a very few years as much too high as they are now too low, and also that trade and commerce and agriculture must go hand in hand. One cannot be depreciated without affecting the other, and if a Corn Bill had passed in 1813, preventing importation when wheat was at 72s. a quarter, some part of the evil we now feel would have been averted; but the low prices have not been produced by importation, as only a fortnight's consumption was imported in the year 1815. The present Corn Bill, with the reduction of taxation, will do everything for the agricultural interest, I should hope, and the less Parliament interferes the better: yet I think it may be wiser to impose heavy taxes on the importation of butter, cheese, ham, hides, skins, and perhaps seeds, although not all seeds."

And again in April, 1816:—

"I received your several favours of the 19th, 27th, and 28th March, respecting the present state of the agricultural interests of the kingdom, and your ideas of the relief to be given, which were so sensible and coincided so entirely with my own opinions upon the subject, that I sent them to Mr. Vansittart, and urged his adoption of them, and you will have the pleasure of seeing that some of them have been adopted. My opinion is that every thing has now been done for the relief of the agriculturist, except the repeal of the tax on the agricultural horses, and on horses used in single-horse carts carrying coal and lime for hire, which is the most unequal and hard ever imposed, and I hope will be repealed, as a friend of mine, Mr. Burrell, has given a notice to that effect the week after the House meets. Mr. Vansittart has proposed a reduction of the duties on horses on small farms, but that is not sufficient in my mind. The other relief to the agriculturist is an extension of time for steeping barley grown in the Northern climates, as the time now allowed is not nearly sufficient for coarse barley, or such as has very thick husks, and which is the case with the Northern barley: if 20 hours more were allotted, it would be a great encouragement to the malting Northern barley, and of course to the grower. I have pressed this upon Mr. Vansittart, but he says that it cannot be allowed without great injury to the Revenue, as frauds would be practiced, and he will not allow that your plan would prevent these frauds: however, I shall again and again urge him on this subject, and also to agree to give up the tax on horses, as I have stated, and then I am sure all reasonable agriculturists would be satisfied. From all the accounts I have had from Lancashire, Yorkshire, &c., &c., and from maltsters, and consumers, I am sure there was no occasion this year to have given up the war tax on malt, but it appeared to have been done in a hurry, and in a pet. A committee (you will see), has been appointed to enquire into the propriety of imposing a duty on the importation of wool, and seeds: of course the committee will receive evidence upon the subject, and recommend what may appear most useful to the community at large; at present, my opinion is that it will be unwise to impose duties on these articles on importation.

"I find agricultural produce rising in all parts of the kingdom, and I perfectly agree with you, and have for many months stated my opinion to every one, that I had no doubt all produce would be as high before April, 1817, as any fair and honest agriculturist could wish it to be, and the less interference by parliament the more beneficial for the agriculturist.

"With respect to my political conduct, I trust and hope that every one of my constituents are well satisfied with it, at least every one whose good opinion I am anxious to have. I look up to Lord Lonsdale, not only as my friend, but as the most valuable, and one of the most liberal, independent, and enlightened of men. I am indebted to him for introducing me to the Freemen of Carlisle, and I do not believe I had six dissenting voices (if I had any) against my last election, and I hope to have as few opponents at the next. I have unremittingly for fourteen years attended to the interests of the whole kingdom, but more particularly to those of the counties of Cumberland and Westmoreland, and have warmly espoused every object which has had in view the improvement of the country, or of the people, and have made it a rule to answer every letter and see every person, and endeavoured to serve every one deserving. I have almost universally supported the Ministry, because I have approved their principles and in general their measures, and when I have differed from them, and from my worthy and most excellent friend (which has only been very seldom), it has been with the greatest pain to myself, and nothing but a conscientious discharge of my duty, which I conceived due to the public as well as myself, could have induced me to differ from them. If I was not a free agent, and perfectly at liberty to act and judge according to my conscience, I certainly would not sit in parliament one day.

"I wish well to every religious and charitable institution, and subscribe as liberally to them all in every place where I have my property or connexion as any of my friends or neighbours, and particularly to the National Schools, and the Society for promoting

Christian Knowledge, as I am more anxious for the support of the Church Establishment than any other."

On the introduction, in 1822, of a Ministerial Bill with a sliding list of duties, Lord Althorp wrote Mr. Bates:—

"My dear Sir,—I certainly do not think the bill proposed by Lord Londonderry is the best measure which could have been adopted for the regulation of the Corn Trade, but I think it so great an improvement upon the present law that I cannot do otherwise than support it. I think the best system under which the farmers of this country could live would be with open ports, a duty of 20s. upon the import of wheat, and a bounty of 15s. upon its export, and other grain in proportion, as I am convinced that the vain attempt to keep up the price of corn by a high import price has been the cause of all our present distress. Without the power of exporting corn in years of super-abundance, there must be frequent recurrence of distress, and the only way in which we can have any chance of exporting is by means of a bounty. I am afraid you will not agree with me in opinion, but it would require a longer letter than I ever wrote (except about cattle), to give you my reasons. I think, however, you will be led to doubt whether I may not be right, if you will consider the subject. These are the grounds: A country that grows its own consumption in average years, must grow more than its consumption in good years, and if it cannot export the superfluity the price must fall seriously low, and the increased quantity cannot make up for the deficiency of price.

"Believe me, my dear Sir,
"Yours most truly,
"Albany, June 1st, 1822." "ALTHORP.

Sir Thomas Lethbridge, M.P. for Somersetshire, in 1822, wrote to Mr. Bates as follows:—

"I am glad to hear you have sent a petition to Mr. Wortley, who, I am afraid however, is not very cordial in our cause. I will ask him on Monday if he has received it, and you may rely on my support. You have now nothing more to do than to send up all the petitions you can to the Lords. I will name some of them, Lords Carnarvon, King, Suffield, Lansdowne, and Holland, and the other noble Lords who are opposed generally to the Government. I should hope and believe they will not only receive it with care, but present, with proper representation, your petition.

"I am, Sir, in great haste, and hourly and daily ever here to serve the cause of the landowners and occupiers, who are now so unjustly used,

"Yours very truly,

"LETHBRIDGE."

Corn and currency questions have now been almost forgotten. I venture to insert part of a letter from a leading M.P., in May, 1822, to Mr. Bates, which truly represents the difficulties his party had to meet:—

"I have to thank you for your letter on the corn bill, which is very much in unison with my own ideas, but, I have unfortunately been prevented, by ill-health, from taking any share in Parliamentary business this session. It is not exactly the influence of the *monied* interest in Parliament that prevents more efficient steps being taken to protect agriculture, as you seem to suppose, but various causes combined. There is a strong party of real agricultural members in the House, but not half-a-dozen of them ever agree on the same plan. Those who are mere theorists and talk of free trade are, of course, united and against every one of the plans of the agriculturists. So are various other interests. Then there are a great many who won't think on the subject at all, or who cannot devote time to it, and, as the simplest way of settling the matter, determine to support whatever the minister proposes. The minister himself when he comes to make his proposition has to consider, not merely what is the *best plan*, but what he can carry through the House, so that perhaps that which is ultimately adopted, may not, even by the minister himself, be deemed the best—and multitudes of agriculturists vote for it, merely because they think it the best they can get. If I am correct in this view, it follows that the great evil is want of union among the agricultural members themselves—one well worth cure. My own opinion is that

Londonderry's bill will pass nearly as it is, but as you allowed me to communicate the contents of your letter, I did so, whenever I received it, to a very intelligent member of the agricultural committee that he might turn them to as much account as he could. In Scotland the breeders are dreadfully ill off; no sales at any price."

Mr. Bates deeply lamented the subsequent free trade legislation, and the prices at his death seemed to justify his forebodings. The result is that the prices of corn, as tested by the tythe averages, have continued much as before the duty was abolished, but the fluctuations in prices have been greater, and the prices of buctcher meat have risen very greatly. The result, however, has been that whole districts in Northumberland have been put into grass, and the population has decreased as the arable culture has disappeared. The Halton Castle estate has been all laid to grass, and let annually in "grass parks," and one man only is employed as cattle herd, where Mr. Bates employed fifteen men, besides women and children, and extra labour in hay time and harvest. The result is the same on Sir William Loraine's, and many other estates in the county.

As a result of this practice I was, during a recent visit to Northumberland, after a very long absence, much impressed with the melancholy appearance of the homesteads. Where I had seen "rich harvests and merry harvest-homes" there is scarcely an inhabitant, and everything is grass. Even the estate of Mr. Grey, at Milfield Hill, has been entirely in grass for some years, and let in grass parks yearly. When I asked for the great improvements made by Mr. Grey on the Greenwich Hospital estate, and in Tyneside, the universal opinion was that the cultivation of the whole district was much inferior to that of 60 years ago. The lands are fast going out of cultivation, and now few herds exist except of Irish cattle, and they of the worst breed that can be imagined, and the sheep on the universal grass are a cross between the Leicester and Cheviot. Few pure breeds exist, and these only in the hands of professional breeders. Many large estates have changed hands, and all have been purchased by gentlemen, strangers in the county, who have made fortunes in manufactures, mining, or commerce. Few resident landowners, who cultivate their own estates, are in existence, and every day the gulph between the landowner and the occupier is increasing.

Mr. Bates was always a large employer of labour, and paid great attention to the welfare of the poor, and objected strongly to the new poor law. He contended that it was quite unnecessary, and that the landowners ought to look after their own poor without official machinery and a large paid staff, and when it was proposed to include Kirklevington in a union with Yarm and Stockton, and parishes mostly in the adjoining county of Durham, he took an active part in the opposition, and insisted that Kirklevington should be comprised in the Stokesley union, in which the adjoining parish of Crathome, an agricultural parish the same as Kirklevington, was included.

Mr. Bates had a long correspondence with the Poor Law Board, and in one of his letters he stated that he had paid rates in eight townships, and never, except once, had a pauper in the workhouse from any of them, and in that case he offered the pauper 4s. a week to keep him out of the house, when he soon after died. The result has been that Kirklevington was included in Stockton union to serve the purposes of the aspiring officials, and the rates, consequently, from being 4d. to 6d. in the pound are now 2s. 4d. or 2s. 6d. This increase of 2s. in the pound on a rental of £1,500 is, it will be perceived, a most serious tax, while the adjoining parish has had a very small increase.

As information on the much vexed question of turnpikes, even at the present day, I may here relate some of Mr. Bates's experience on this subject. The road which passed Kirklevington had originally been a country township road to Yarm, Stockton, and Northallerton; in the other direction there was amply sufficient for the estate. An act, however, was obtained to make this into a turnpike, and by making about five miles of new road from the south, to make a more direct road for the thorough traffic from the towns on the east coast to Leeds and London. The first act provided that there should be no tolls between Kirklevington and Yarm. When it was sought to renew the act this exemption was disregarded, and Mr. Bates considered it an attempt at imposition, and a breach of the faith on which the first act had been obtained, and he gave it his opposition. All, however, that he could accomplish was, even when he opposed the bill in the House of Lords, that the agricultural produce of Kirklevington township should pass toll-free, at gates in the township. A railway nearly alongside this road has taken off all except the local traffic. The townships were called on to keep the road in repair, and the township surveyor was appointed by the magistrates, without the consent of the ratepayers. To this Mr. Bates objected, and he objected to the surveyor purchasing the road materials from one of the magistrates, and several appeals were made to the quarter sessions on the subject. The agitation of the subject resulted in a change in the law, whereby the ratepayers, and not the magistrates, appointed the road surveyors.

Soon after the death of Mr. Bates, a new act was obtained at a cost of £800, which relieved the towns at the extremities of this road from tolls, but took away exemptions from the country townships: the result being that Kirklevington and other townships have to pay all tolls, even on the road materials, and to keep the road in repair, and the tolls are applied really in repaying the money borrowed to pay the expenses of the acts, and the salaries of treasurer, clerk, and surveyor.

The improvement of the wastes and morasses in Northumberland, mentioned by Mr. Trevelyan in his letters, occupied much of the attention of Mr. Bates. Large tracts of such land exist near to his Broadpool farm. His father was the first person who began to improve such land. He had purchased about 600 acres of land mostly morass, which was sold to defray the cost of the Act and Enclosure of the Forest of Lowes. The purchase money was £400, and the conveyance cost only 4d. The Act of Enclosure provided that the receipt of the Commissioners for the purchase money should vest the estate in the purchaser. This land was drained and limed, and plantations made for shelter. There for many years a very fine herd of West Highland cattle was kept.

Mr. Welles, in the first volume of the Journal of the

Royal Agricultural Society, 1840, page 348, gives an account of the kyloes he bought from this land, and from which he established a herd in Herefordshire, and found them much superior to what were obtained from the Highlands of Scotland. Mr. Bates wrote much in the Agriculturist publications, calling the attention of the owners to the importance, if only for shelter for stock, of planting much of such high lands. The Duke of Northumberland and the Commissioners of the Greenwich Hospital, on the division of the commons in Corbridge parish, each planted about 1,000 acres, which is now very valuable timber, and in many places the grazing is now more valuable than before the land was planted.

Several Swedish gentlemen, who have visited Northumberland on agricultural affairs, expressed their opinion very strongly on the fact that we imported large quantities of timber and allowed our own hills to remain naked. Mr. Bates prevailed on some of his friends in the County of Durham to plant, and they soon felt the benefit, both in the shelter to stock and the sale of timber, even when very small, for glass-house crates.

Mr. Bates had bestowed great attention on the agricultural machinery and the improvement of implements of all kinds, and thrashing machines. He was also familiar with the attempts made by the Society of Arts to obtain a reaping machine, and the premiums that had been offered for such an implement.* Mr. Bates had

---

* See transactions, Vol. I. Among the Vice-Presidents were Hugh, Duke of Northumberland and his son Hugh, Earl Percy, and premiums were offered in 1783, for 1.— The course of crops on strong land. 2.—For a machine to answer the purpose of

heard from his American friends that such machines were very common in America. I insert an extract to show what he did to introduce the implement into England. "The General Annual Meeting of the Yorkshire Agricultural Society was held August 30th, 1838. Earl Spencer presided, and after a few preliminary observations, begged to ask, 'had any member any observations to offer to the meeting?' Mr. Bates, of Kirklevington, observed that a reaping machine had been invented in America, and he thought it would be the duty of such a society as the present to send an engineer to examine it. He had heard it highly spoken of, and believed it was a highly valuable implement. Earl Spencer thought that before taking such a step some preliminary enquiries should be made. Mr. Bates said that a friend of his in the State of Illinois had seen it for three years, and bore the fullest testimony to its utility. Mr. H. S. Thompson, Kirkby Hall, said that he had seen an implement of the kind in Scotland, and thought it would be as well to examine that before they sent to America. Mr. Bates said that probably they were both of the same kind." Nothing was done by the Society, although a trifle of the money given in prizes

reaping or mowing corn, by which it may be done more expeditiously and cheaper than any method now practiced, provided it does not shed the corn more than the methods in common practice, and that it left the straw in such manner as it may be easily gathered up for binding. 3. –For a method of destroying or burning the smoke of fires belonging to steam engines, &c. 4.—To the planters of any of the British Islands in the West Indies who shall impress oil from the seed of cotton, and make from the remaining seed hard and dry cakes as food for cattle. There were also various premiums and prizes for planting oak, ash, and other trees, &c.

would have tested the merits of the machine, and it is a curious and perhaps a melancholy fact, that a machine which had been invented and used in Scotland many years before thus brought to the notice of Earl Spencer and the and the Yorkshire Society, should be passed over in silence for 13 years until it was brought into notice by the Great International Exhibition of 1851.

Mr. Mackenzie was the only pupil, if he might be called one, that Mr. Bates ever had. Mr. Bates declined to take pupils, but many young gentlemen made prolonged visits at Halton Castle, to see whatever could be of interest or use to them. James Fawcett, Esq., of Scaleby Castle, Cumberland, and his cousin Mr. John Vitre, now deceased, spent a considerable time at Halton Castle. Mr. Fawcett was a great favourite, and much esteemed by Mr. Bates, and by his artistic talents preserved the recollections of many of the animals there. Mr. Fawcett's portraits of the Duchess, of 1804, her son Ketton 1st, Duchess 1st (1810), and Comet, I have lately seen, and they convey a good representation of what an improved short-horn ought to be. Mr. Bates offered Mr. Fawcett, a Duchess heifer as a present. Mr. Fawcett, in common with all visitors to Halton, had a most lively recollection of the pleasant time spent there, and a proper appreciation of the character of Mr. Bates. Many made visits, or rather pilgrimages, to Halton to see the old place, and they never wearied in telling of the incidents of their intercourse with Mr. Bates. I insert the following, from a memorandum of Mr. Fawcett's:—

"It is pleasant to recall to mind the memories of by-gone days, in the sunny period of life, the gleams of which are amongst the most pleasurable sensations of which we are capable. With this view I determined, a few years ago, to indulge myself with a re-visit to Halton Castle, the scene of many happy hours spent with my old and much valued friend Mr. Bates. With much interest I again beheld the old place, somewhat altered, though in many respects, much as it was in days of yore. There were the old ash trees, and the paddock and shed where the elephantian Ketton the first, and his successors, stalked about; and the rich pastures to the North, where the renowned herd roamed in luxurious plenty. The Old Chapel. The Border Tower, and appended Mansion, and in it the great room, where, enclosed within the precincts of the large old fashioned screen, with a blazing fire in front, and copious libations of tea, with old Barbara's cream and currant cake, we cosely read and discussed Davy, Marshall, Culley, &c., and also short-horns, and everybody and thing connected with these, till the small hours. The days were spent in looking over the farm operations, visiting the pastures, and riding out to see the stock and farms in the neighbourhood. On one of these occasions we visited Mr. Donkin, of Sandhoe, the purchaser, from Mr. Bates, a few years before of the famous Daisy cow (then called Duchess), grandam of Duchess 1st, she was then said, I think, to be in her twentieth year, and not having bred for some time, was about to be sent to the butcher. I was so much struck with her fine clear eye, extraordinary breast and shoulders, and general appearance of an 'old has been,' that a drawing was made of her. On another occasion our ride was extended to the Tees Valley, and the principal herds of that famed district were visited, and among them, that of Mr. R. Colling, at Barmpton, which was shortly after disposed of. We spent a pleasant day with that gentleman, and there was much interesting discussion about them. He pointed out to us a favourite four year old cow, which had the week before produced him three living calves at a birth, having been previously barren. This he attributed to the effect of ploughing on the farm the foregoing spring. We also saw Mr. C. Colling, then living at Croft. Mr. Bates's criticism on these herds was by no means sparing where he could see faults.

"At this time Mr. Bates had several other good tribes besides

Duchesses, and a number of beautiful cows of more or less affinity to Argyleshire heifers, which he had procured from thence at considerable cost and trouble. The produce of these bred back to the short-horn, seemed to answer admirably, and after a few crosses had all the appearances of first-class short-horns; but whether from the diminished size of his farm at Ridley Hall, or from the gradual increase of the Duchesses and Cambridges, none of them appear to have been taken to Kirklevington.

"The character of the Duchesses at this time was that of good and handsome wide spread cows, with broad backs, projecting loins and ribs, short legs, prominent bosoms. The head was generally rather inclined to be short and wide than long and narrow, with full clear eyes and muzzles, the ears rather large and hairy. The horns of considerable length, but of free waxy quality. They were good milkers, and, for the most part, had a robust healthy appearance. The colour almost uniformly rich red, with, in many of them, a tendency to white about the flank. They had also generally what Mr. Bates called the Duchess spot of white above the nostril; but this peculiarity does not seem to have descended to the tribe of the present day, though it appears in the roan introduced into them by the Oxford cross.

"Mr. Bates, being an intimate family friend, was often at Scaleby, and induced my father to try a cross of the new breed in his long-horns, and for this purpose insisted in furnishing us with a Ketton bull, which answered so well that no more long-horns were got, and by the purchase of a few well bred short-horns, many of them recommended by him, and some from his own herd, a complete transformation was effected. Amongst them were some descendants of the cow, Princess, whose daughter Elvira was brought into Cumberland, and produced some good animals. (Mr. Fawcett's drawing of the old cow in her fifteenth year is given, shewing the style of that tribe at the time.) She was sold at the Wynyard sale, for 96 guineas, her sisters, Anne Boleyn and Nell Gwynne, not fetching quite so much. To revert to the Duchesses, a strange anomaly occurred in the case of the sixth. I recollect her being calved, she was very handsome, and of the most orthodox colour, but with a round spot, of several inches, on the flank, of the deepest black. Whether this indicated a harking back to some ancestral highland alloy, or freak of the cow's imagination, is a curious question.* This long and, I fear, trivial and personal letter requires apology, but it was thought some of the details relating to a departed one, so eminent as Mr. Bates, ought to prove interesting to the lovers of short-horns."

PART VII.

Mr. Ramsey's Tyneside Herd—Mr. Bates's First Bulls and other interesting matters—The Ketton Herd and Mr. C. Colling's anxiety—The Barmpton Herd and Mr. Bates on a Pure Love of Money—Mr. Bates at Home.

George Hepple Ramsey, Esq., of Derwent Villa, Winlaton, Durham, probably the oldest agriculturist now left, and an active member of the Tyneside Club, was a frequent visitor at Aydon Castle, and at Halton. He says the Aydon Castle cows were considered the best in the district, and were very fine and large cows. He was at Aydon Castle once after the purchase of Duchess 1st and Mr. George Bates ridiculed the purchase, saying he had many much superior in his own herd, which he should have had for nothing. Mr Ramsey says that the Aydon Castle cows were much larger than those at Halton, but that certainly they did not show the breeding of the Duchess, and that they considered Duchess 1st a "shabby" animal. Mr. Ramsey relates the care Mr. Bates took to point out to young agriculturists, who went in great numbers to view the

* A similiar event occurred within my own knowledge. A pure bred short-horn cow rubbed against some black cart grease, and made a black spot on her skin, and produced a calf with nearly the same black spot on the same place.

stock at Halton, the properties cattle ought to have; and he frequently had them blindfolded, and made them handle the cattle, that he might be assured they understood and appreciated the good qualities and judged properly of the different animals. I have endeavoured to obtain information from the few persons now living who could recollect the herd of Mr. Bates at Halton Castle, and perhaps the following is worthy of a place as showing what the impression of the writer was. He resided near Mr. Bates and afterwards settled in Essex as a successful agriculturist and feeder of stock. He writes:—

"The first bulls I remember Mr. Bates using were Daisy Bull, Styford, Laird, and Chieftian, until that emperor of all bulls, Ketton, came into use, and it was with telling effect. Tommy Thompson, the cowman, said that he never got a middling calf all the eleven years he was in service, and I think I can see the grand old animal standing in the bull park with his fine head, placid countenance, his beautifully arched neck, his deep and roomy chest, his short and wide-spread legs, his handsome shoulders and full crops, his long, straight, and level back, his heavy flank and deep ribs, his well-formed beautiful quarters and heavy thighs, and his tail so nicely set as to give symmetry to the whole form. How oft on my youthful mind was impressed the idea that I should never see his like again, for his image was so imprinted on my memory that whenever I began to examine a prize bull, Ketton came full in view, and then many defects were soon prominent. Mr. Bates' herd soon began to develop its superiority, and the jealousy began against him because he would not allow any cows but his own to be put to Ketton, and he did not rear many bulls, except for his own use. There was one, I remember, the very image of Ketton that Lord Althorp (Earl Spencer) bought. A grand animal, so like his father, with that fine fleecy silky coat that so remarkably distinguishes the Duchess tribe. That in a great measure kept the Duchesses in Mr. Bates' own hands for so many years. Although

Mr. Bates used Ketton for so many years, yet a Duchess heifer or bullock could easily be picked out of his herd; there was something in their very countenance, and in their prominent gait, and, above all, in their superiour *touch* like none else, for *in that quality they had no equals.* After Mr. Bates left Halton I had not the opportunity of seeing his herd so often, but the last time I was at Ridley Hall I was highly gratified at seeing second Hubback, a bull well worthy of that distinguished name, and fully came up to my reality of what a bull ought to be. I compared him in imagination with Ketton; and, if prejudice had not been deeply seated, I might have said he was as good as Ketton. Mr. Bates smiled when I told him that he was surely next to Ketton; even he would not say that he was better than Ketton. Then Belvedere was the next grand animal that I exactly liked. I had every opportunity of seeing Jobling's, Wailes' (of Bearl), Charlton's (of Bearl), and the Gibson's, Angus' and Tyneside celebrities; but as these were mostly of the Mason tribe they did not take my fancy, so much was I tainted with deep prejudice. I said that, if I was anything of a prophet, I thought Mr. Bates's breed would be had in remembrance when theirs were entirely forgotten. And then came that Duke of Dukes, and the bull of bulls, with his never-to-be-forgotten dam, Duchess 34th. Her equal I never saw, nor do I ever again expect to see so grand a cow. I don't think I need say anything of Mr. Bates's character, except that he was an enthusiast regarding short-horns, being a very superior judge of a good animal. He would never praise any that were in his opinion below the mark, but passed them by in silence, so that many said he was so very prejudiced. But experience proves that it was far-seeing judgment. Mr. Bates took little interest in politics, except in the corn law question, but as regarded the breeding of cattle and sheep, he felt a keen desire to know which were the best and most profitable to the farmer, and which would supply the most and the best quality to the consumer at the same cost: and he came to the conclusion, that well-bred short-horn cattle, and the Leceister sheep, were decidedly the best for producing both food and raiment; and surely to no other man is the country so much indebted as to Mr. Bates, not even to the Messrs. Colling, as it was by Mr. Bates purchasing the dam of Ketton, for 100 guineas (a price that was considered incredulous), that brought their stock into such repute. Every newspaper was trumpeting the marvellous wonder, and every gentleman who wanted to commence improving his herd went to the Collings for a bull, so that it was considered a great favour to get one from them. Previous to that Mr. Mason and the Joblings were considered the great men, but that was immediately altered, and every one that talked of improved short-horns, set their affections on those of the Collings. But still Mr. Mason had a party that stood up for his breed very stifly, but the 'dons' who were able and willing to give high prices flocked to the Collings, and if they could be served there, well, if not then they must fall back on Mason or Jobling. Then came Mr. C. Colling's sale, which produced such a sensation throughout the whole kingdom, Comet making 1000 guineas, and Duchess 1st 183 guineas. Then people began to think what a judge Mr. Bates must have been to see such extraordinary merits in his 100 guinea cow. And if he would have sold his young bulls, by Ketton, he need not have made steers of any of them. Only his particular favourites enjoyed the privilege of a Duchess bull."

I have often been informed that the Ketton herd did degenerate after 1804, and that it caused Mr. C. Colling great anxiety, and he dreaded that the fame of the herd would soon come to an end, and this was the real cause of the sale in 1810. The particulars of the sale, although so well known, I will refer to hereafter. Short-horn breeders may investigate it to see how far any of them can now be traced, and whether Mr. Bates was not correct in saying that there are not any descendants of the females except the Duchess that can be traced. None of the stock went to any great distance. Three bulls and the Duchess went into Northumberland, being more than went to any county except Yorkshire. As Mr. Bates stated, he did not purchase or wish to purchase any animal except Duchess 1st.

The next great sale was that of the Barmpton herd, Mr. R. Colling's, in 1818, and Mr Bates did not buy anything, and the only animal he thought worth having was the Marske bull. The prices were such as again astonished the world. I shall also notice this sale afterwards.

Mr. Bates used to say, and indeed truly, that money-making was the ruling motive that governed the breeders of short-horn cattle, and not the improvement of the short-horns. Had the latter been the *main object*, the short-horn cattle might have been of double the value they then were in these kingdoms, and spread over the whole habitable globe, rendering an equal improvement wherever they went, to the cattle of our younger brethren in the United States of America, as well as in English possessions.

Mr. Bates might have kept all his calves for bulls and sold them at high prices. After the advent of Belvedere, he kept more calves for bulls, but very frequently not by any means the best looking or most promising of the stock, but he frequently had ten or more bulls in his possession which he never used and had no desire to part with except when he considered they would improve the herds into which they were going. As an instance of the principle which guided Mr. Bates, I may mention that I well recollect the late Mr. Wetherell, the well-known breeder and judge of short-horns, coming to Kirklevington. Mr. Wetherell—the "Nestor of short-horns," justly called—was one of the purchasers of Comet, and had bred many excellent cattle, and always looked to the constitution of his stock. Mr. Wetherell did an extensive business in buying and selling stock, and no one knew the merits of short-horns better. He asked Mr. Bates if he had any bulls for sale. Mr. Bates replied that he had, and Mr. Wetherell rejoined that he wished to buy one or two, and after viewing the stock with Mr. Bates, Mr. Wetherell selected two which Mr. Bates said he would sell him. They went into the house and had dinner. Mr. Bates then enquired about the herd into which Mr. Wetherell proposed to send the bulls. Mr. Wetherell asked in reply, " Of what consequence was that as long as he got the money for them?" Mr. Bates rejoined, he "*would not sell any man a bull unless he knew the herd to which it was going*, for if the cross did not answer *all the blame would be attributed to the bull.*" I attended Mr. Wetherell, on his leaving, to his horse, and he could not refrain from expressing his opinion in strong terms in regard to Mr. Bates, for refusing to sell his cattle at high prices so long as he got paid for them.

When at home alone, Mr. Bates read a great deal, mostly on agricultural subjects and political economy. But he had a most extensive correspondence to keep up. When disengaged he regularly walked out and visited all his animals, and spoke and talked to them as if they understood him. In the field they immediately came round him, and he generally caressed them by rubbing his hand down the under side of their necks, and so accustomed were they to it that those he did not immediately notice would poke him with their horns from behind, and so call his attention to them, while others would take hold of his coat tails with their teeth and pull at him. To get quite away from them he was obliged to walk very fast, and even then some of the younger ones would frisk and scamper after him. His shoes and clothing had often to be changed, as they were so impregnated with the smell of the cattle that it was not difficult to tell where he had been. Mr. Bates would sometimes tell Robert Thompson, the cowman, that he would help him to drive the cows to the field, and while doing so the cows would come round and not leave him, on which Robert, becoming irritated, would exclaim " I wish you'd keep out oft' way, you do far mair ill than good, for they wont leave you, and ther's no driving them."

Mr. Bates carried his simplicity of living into all his household arrangements, for he seemed to have had no idea of even ordinary accommodation. Although at Halton and Ridley Hall, he had good mansions, yet at Kirklevington he never thought of ordinary accommodation, until his friends remonstrated with him, and he had the rooms fitted up. The dining room was perhaps the smallest room of its sort in the county. But never did so small an apartment contain so many celebrities—Bishops, Peers, and Agriculturists of all classes. Although Mr. Bates had only a housekeeper and housemaid, yet, the most fastidious could not object to the style of the entertaiments he gave or his hospitality. The Duchess of Leeds, who preceded Lady Pigot as an admirer of short-horns, was often a visitor, and discussed their merits, and many letters passed between them on the subject. Mr. Bates had the same simplicity in the accommodation of his stock. The buildings at Halton were poor and detached, in consequence of the occupation having been in separate holdings. And at Kirklevington he never improved on the original buildings, the additional accommodation was of the most ordinary and temporary description. The best cattle were kept in cow sheds, with sides made up of whins or gorse, and the roof of red tiles, not pointed with lime; one side had long been kept open. Mr. Bates had constantly seen the large and superior buildings on the Northumberland farms, but had no idea of imitating them. The buildings erected by Bailey and Culley are models to this day in Tweedside; but Mr. Bates was as free from the vanity of fine buildings, as he was of personal appearance. There was not even a decent calf house on the place, and at last I got him to have one erected after the model of that of Mr. Hutchinson, of Grassey Nook.

The dress of Mr. Bates, like Arthur Blayney's, was plain and studiously neat and becoming, although it was of the old form and fashion. He was temperate and abstemious in his habits, and he was generally almost insensible to both eating and drinking, while his food was ever of the simplest discription. His housekeeper got so accustomed to his ways that she used to watch him and take care that he had his breakfast before he went out in the morning. It frequently happened, however, that he got out without his breakfast, and that he did not return until mid-day, and then the housekeeper would ask him which he would have, his breakfast or his dinner, and he would exclaim, "breakfast; have I not had any breakfast?" He would take his breakfast, generally of oatmeal porridge and milk, and then take his dinner, as he usually did, at twelve o'clock, when without visitors. When alone he never tasted anything except water, but with company he always joined in wine; spirits he never tasted and malt liquors only once to my knowledge.

Mr. Bates had a great dislike to any outward show; "He could never be persuaded to keep a carriage and very seldom hired one," making an excuse that it was too great a luxury to indulge in, and that neither his father nor his grandfather ever kept such a thing, and he was no better than either of them, and riding on horseback he said was much healthier. A Mr. Foster, of Carlisle, made him a present of a strong pony to ride, but it went blind soon after he got it, yet he continued to use it for many years afterwards. When he first went to Halton Castle, he owned a very excellent mare, which he rode for several years, and she sometimes carried him a hundred miles a day, when he wished to attend markets many miles apart. He was at times put to some inconvenience for want of a conveyance. On one occasion he hired a dog-cart to convey a friend who had visited him to Mr. Maynard's, of Harlsey. They had not proceeded half-a-mile from home when the mare shied at some Irish people resting by the wayside, when Mr. Bates, who was earnestly engaged in talking and not minding his driving, suddenly pulled at the reins, in doing which they got under her tail and she began to kick and broke the shaft, and turned the carriage completely over. Mr. Bates was much bruised and could proceed no further, and his friend and baggage were sent on to Harlsey in a cart.

Mr. Bates had a fine breed of Cleveland and road horses. The Cleveland breed is now nearly extinct, having been crossed so much by blood horses. Mr. Bates bred many fine horses, but having lost two fine young ones by their falling into a quarry, he never afterwards took any interest in horses. He never joined in field sports. When young he had shot each season, but having met with an accident from a large gun bursting in his hands, he never took one up again. This large gun was purchased to shoot the ravens, which carried off the lambs, and I believe they were the last seen in the County of Northumberland.

Mr. Bates, like Mr. Curwen, took great interest in the welfare of the poor, and in their education, and his religious feelings brought him into intercourse with what are called the Evangelical or Low Church party, who then looked to the Rev. Charles Simeon, at Cambridge, as their leader. Mr. Bates maintained a constant correspondence with Mr. Simeon, the Rev. Leigh Richmond, and other members of that party, both in Cambridge and in the country, especially in Cumberland, with the Carus Wilson family, and he was a liberal contributor when calls were made upon him to assist the cause. He contributed largely and took a very active part in the establishment of the British and Foreign Bible Society, both in his own neighbourhood and in Tweedside. Mr. John Grey made his first public speech in support of this society. He took much interest in Sunday and Infant Schools, and he incurred great unpopularity by endeavouring to stop drinking on Sundays at the public houses in his neighbourhood. He was also intimate with Mr. Wm. Wilberforce. The state of the Church was a constant cause of regret to him, and as there was service only twice a month in the Halton chapel, he endeavoured to have a service on the alternate Sundays in the chapel, but could not obtain the Bishop's consent and was obliged to hold it in the castle, where he frequently had a clerical friend to conduct it.

I venture here to insert a letter from the respected Vicar of the Parish to Mr. Bates, then in Edinburgh. The Vicar had lost one son in the Navy, the other was severely wounded under Abercromby in Egypt, and became a distinguished soldier. The notice of the pens may now be interesting :—

"Corbridge, Newcastle-on-Tyne,
"Wednesday, March 21st, 1810.

"Dear Sir,—It has long been my intention to send you a few lines, but something or other has always intervened to prevent me. I was much obliged for your letter in answer to mine enclosing Mr. Gray's, a gentleman whom I know has no small curiosity to see the improved breed of stock in the North of England, and I think it is very possible he will one day or other beat up your quarters at Halton Castle. From your various pursuits in Edinburgh, you cannot have much leisure time upon your hands, yet I am willing to believe that you will find a few minutes to spare for the perusal of what comes from an acquaintance in this part of the country. Though you have been removed this winter, perhaps a degree and a-half to the northward of this latitude, I do not apprehend you have felt a colder climate, than what you might experience upon Halton Downs. Sunday, February 25th, was one of the severest days I ever travelled in to the chapel, and even scared my friend Dobinson, of Great Whittington, from attendance. My congregation was consequently chiefly composed of your own township. Last Sunday was almost as cold though a westerly wind, but the men of Whittington ventured to face it as well as a lady to be churched. Upon my return home, I called at Aydon Castle, and was happy to find both the master and mistress of the mansion in good health. They were preparing for dinner upon a roast turkey, to partake of which I received a most cordial invitation; and had not my afternoon duty stood in the way, I would have 'devil'd' for them both the rump and the gizzard, for the keenness of the air had set me very sharp. Your father said he thought he had caught a little cold, but as to your mother she looks like a four-year-old, and I'll be bound to say, notwithstanding her years, there is not a fairer or finer skinned woman in all Edinburgh. Mr. Bell informed me everything was going on to his satisfaction, though a rascally fox had been paying you a visit more than once, but had committed no depredations. It is supposed that he had even the audacity to take up his abode in your garden, for he had been paddling upon the top of the hen house, and had leapt over the wicket that leads towards the chapel. He had also attempted a burglary upon a house where you had some lambs, but found it too well secured, in short he has hitherto been baffled in all his efforts, but it is necessary to be upon guard against him and shut up the troops early in the evening. Soon after you left the country, Mr. Giles sent us down three turkeys, a present I understand to Juliana, which she will be happy, I am sure, to thank you for when she sees you. Unfortunately they all proved hens, I took the liberty therefore, for my daughters have been absent all the winter, of placing one of them upon the spit, and have since sent the other two to see their old gentleman acquaintance again at Halton. After remaining there nearly a week, they returned on Saturday, and I suppose everything is now as it should be. Mrs. Wilson and I have spent our winter alone. My daughter Jane is at Bath. Kate in London, Elizabeth with Mr. Headlam at Wycliffe, Juliana at Richmond with Mr. Cornforth, and my son fighting the battles of his country in Portugal. His last letter was dated about the middle of January, when he was in good health and spirits. I begin to think that he will soon see some serious service there, and have more fears for him than I am sure he has for himself. I know he will do his duty and I pray God he may return in safety. Yesterday I called upon Miss Donkin at Sandhoe, where I saw Mr. Donkin who gave a good account of his wife. Henry has returned from town and had seen Mr. Richardson, who was perfectly well and as pleasant and entertaining as usual. Miss Donkin has received a number of the patent pens, any of which, if you choose, you must mention soon, as I understand there are many applicants for them. I have taken three of them, each different as to its mountings; the particulars concerning them and the prices are as follows (the prices are somewhat higher than it was supposed they would be):—

	£	s.	d.
The Silver Handles	0	8	0
Wood Handles, silver mounted	0	5	0
Wood Handles, brass mounted	0	2	6
Brass Tube Handles, with Pencil	0	4	0

"To one handle you may have as many nibs to fasten on as you please, the brass nibs at 2s., the silver at 3s. You may choose a hard nib, a soft nib, or a middling one, I am now writing with one.

"Mr. Stokoe was here last week. He and the family are all well. Tell William if he can learn how to keep old women alive as long as his father has done Mrs. Glazonby under similar circumstances, he will never want practice. I must beg you to present Mrs. Wilson's and my best respects to Mrs. Rutherford and the family, any of whom we shall be at all times happy to see, and do not forget us to my aunt Coby. Mrs. Wilson joins with me also in best compliments to Mr. William Stokoe and yourself, and I remain,

"Dear Sir, yours most truly,

"GEORGE WILSON.

"P.S.—To your late friend Willy Jobling I have not heard that any successor as bailiff has yet been appointed, but I understand Wailes' brother is the most likely to succeed him. His farm is to be let, and Willy Hunter and Michael Brown, jun., that married Miss Hall, are talked of as tenants."

The scandal in the church became so notorious that he appealed to the Bishop of Durham to interfere in the case of a clergyman who was habitually incapable of performing the duty on Sundays, and in another case the friends of the church felt compelled to interfere, and they prevailed on Mr. Bates to undertake the management of an action against a clergyman for the seduction of a servant. The action was defeated, however, by the objection that the servant was not in her father's house, and therefore he did not lose her services. In the first case, the parishioners almost resented the attack on their minister, and in the other the parishioners presented a silver cup to the clergyman to celebrate his triumph at the assizes.

Before entering into any statements respecting pedigrees and the alloy or cross with the polled Scotch cattle, I may remark that I have invariably found the statements of Mr. Bates correct in the minutest details. But at the time I had great difficulty in understanding many things mentioned by him. He always spoke of the Messrs. Colling "admitting" or "acknowledging" to him, but when I learnt the character and the habits of the Messrs. Colling, as mentioned hereafter, the matter becomes clear enough. Mr. Bates only got his information by stating the facts, and then asking if he was accurate, and then they only admitted or acknowledged that he had made correct statements. The facts about their pedigrees, and the supposed mystery about the alloy, are thus stated by a writer conversant with the subject.*

"It may be here not unnaturally asked, how came it to pass that this subject had become a matter of conjecture and of controversy? Was nothing known as to the system the Collings had pursued? What where the facts? The brothers Collings appear to have been retired, reserved men, little thinking that their cattle breeding proceedings were to become matter of public curiosity, or were to provoke controversy. They kept no record of their proceedings while they were in operation, nor did they leave any at the last; and seeing, as we have said, that the interests, or supposed interests, of various parties conspired to make it expedient to mystify the facts, it is not surprising that they were so mystified to a great extent. Now, however, that this mystification is passing away, the facts are, in our opinion, abundantly obvious."

* Letter in the *Mark Lane Express*, 12th April, 1858, by *Dunelmensis*.

PART VIII.

The Alloy Blood fully discussed—The "Booth Men" and Mr. Thornton's *Circular*—Mr. Carr not a good advocate, but denounces his own theory—The general custom and value of "Handling"—Mr. Bates on in-and-in Breeding.

The alloy blood was a stain in the pedigree of short-horns that Mr. Bates would not hear mentioned. He considered it a disgrace to the breed. Much that has been said and written about its history is, I believe, now made clear. Mr. Bates was well aware of the pedigrees of all the cattle of the Messrs. Colling, and knew what was meant by grandson of Bolingbroke and Lady. But the subject of the alloy was only cleared up about 1820. I have inserted the results of the sales of 1810 and 1819 of the Colling's in the appendix, to which I shall refer my readers.

There had always been an impression that the excellence of the Colling cattle arose from crosses with Scotch cattle, and the West Highlander was usually mentioned as the cross, and Hubback was generally supposed to have had such a cross. Sir John Sinclair had always been an advocate for the excellence of the native Scotch breeds and urged the Scotch farmers to pay attention to their breeding and to improve them. Of Mr. R. Colling's sale in 1819, Sir John Sinclair expressed his opinion very freely that the prices obtained at this sale were unwarranted by the intrinsic value of the stock, and that they had no extraordinary value above that of the Scottish breeds, and assumed that the Colling cattle were only crosses, such as Mr. Bates had bred with so much success in 1808. The short-horn breeders of course took up the cudgels with Sir John.

The constant reference of Mr. Bates to this alloy was a great source of imitation to the owners of such blood. The purchasers of such animals did not like to say much on the subject as a discussion would only depreciate the value of the stock they had got. But few of them had the least idea that the grandam of Bolingbroke was anything except a pure bred animal from such a grandsire. George Coates had purchased the Galloway bred cow from Mr. Smurthwaite, for Colonel O'Callaghan's dairy. They were not selected as the best of their class, red Galloway cattle being always considered worse grazers than the black, and after attention was paid to the quality of Scotch cattle, the red breed soon disappeared and black only have been kept.

It was fortunate for the alloy cattle that the cow's red colour could not by cropping out in her descendants betray their Scotch origin; and although hornless ones might be bred, certainly the horns were often of the very shortest and very different from those of the Duchess tribe. As Lord Althorp says of Mr. Champion's cattle, they might be very good, but they are not short-horns, much less improved short-horns: and Mr. Bates always spoke of this alloy with the greatest abhorrence. He could not comprehend how C. Colling could admit such blood into his stock, and that such a cross should be bred from the dam of Favourite. Their short-horns had certainly not then deteriorated in size or constitution, as they afterwards bred the Wonderful Ox, and the Roan heifer, the White heifer and Comet. I have stated in the appendix the pedigrees of this alloy cross.

I had written more fully on this subject, but as it has become almost an article of faith with the "Booth men" to support this cross or alloy, I think it better to copy what other and independent persons have said and written on the subject, and I cannot do it more appropriately or concisely than by giving the following extracts from the writings of one who has recently devoted much research to the subject.* In his account of the Ketton sale, Mr. Thornton says:—

"The sale was on a fine October day, and early in the morning people rode and drove to Ketton, leaving their horses and gigs at the adjoining farms; all the strawyards were full, and the throng at the sale immense; everything was eaten up, so that bread had to be sent for into Darlington. Mr. Kingston, the auctioneer, sold the cattle by the sand-glass, and in accordance with the custom of the time, received about five guineas for the business, the work of the sale falling more on the owner than the auctioneer. The Rev. Henry Berry writing in 1840, says, 'that in Charles Colling's sale in 1810, there were some very fair lots bred with a kyloe cross, and he, C. Colling, publicly declared such to be the fact.' Eye-witnesses, however, believe that no such statement was made to the whole of the company, for no general knowledge of the alloy blood prevailed, and it is probable was unknown, save to Mr. R. Colling and Mr. Mason, and one or two others. The cattle were not fed up for the sale, but kept naturally, and sold when they were in great condition from natural keep.

"Soon after the excitement concerning the extraordinary result of the sale had abated, Sir John Sinclair, then one of the greatest agricultural authorities of the day, published a paper depreciating it, and stating that the pure bred short-horn was a mixture of other breeds. This excited some controversy. Sir John Sinclair visited R. Colling and Mr. Mason, whereupon he wrote a second paper withdrawing some of the assertions made in the first. Mr. Mason

* Thornton's Circular, 1869, volume 8, page 238.

in the meantime gave an account of the Galloway cross or alloy blood; wherein he stated, 'That he did not recollect any experienced breeder who made an offer for the mixed breed, and he was sure that if Charles Colling had not made that mistake, his stock at Ketton would have sold for some thousand pounds more.'

"This was read by Colonel Mellish, at the King's Head, Darlington, and caused great consternation in the neighbourhood, as the catalogue did not mention any particulars of the breeding of grandson of Bolingbroke (280). Many were dissatisfied, and others said if they had known of the transaction, they would not have purchased. Mr. Robert Colling also told Mr. Wiley, that he had no doubt it was quite a thousand pounds loss to his brother, he having the alloy blood in his herd. This, eleven years afterwards, became with the breeding of Hubback, the subject of a great controversy in the *Farmers' Journal*, and continued more than a year, with the Rev. Henry Berry and Mr. H. Cottrell on one side, and Major Rudd, Mr. John Rooke, and Mr. John Hutchinson on the other.

"There is no mention of Charles Colling, who was alive at the time, coming publicly forward to refute the various statements, but in a private letter to the Rev. Henry Berry, he stated 'that Hutchinson was egregiously wrong in charging the Collings with an indiscriminate use of kyloe blood.' George Coates declared unequivocally that he never observed anything in that stock designated pure short-horn, that could induce him for a moment to entertain a suspicion that the animals were nearly, or remotely, allied to the kyloes. Mr. Charge, as well as Mr. Coates, and Charles Colling, always deemed Hubback a pure short-horn; and neither he, nor his descendants, when put on cows of the pure blood, begat any calves which denoted, in their features or colour, any other breed than the pure short-horn.

"His (Charles Collings) stock had capacious chests, prominent bosoms, thick mossy coats, mellow skins, with a great deal of fine flesh spread equally over the whole carcase, and were either red and white, yellow roans or white. The produce of the alloy blood increased in size, rotundy, and heavy flesh, but afterwards seemed to lose thei fine hair and milking properties. The highest priced cows at the sale were those in the highest condition, and they mostly of the alloy blood."

Pilot, Mr. Booth's bull, is the representation of this alloy stock. It does not appear that Mr. Booth made any complaint on the subject. The results of breeding from Scotch, Irish, Welsh, and other cows, and the fatal effects of introducing their blood into the short-horn herds, is so eloquently and conclusively stated in a recent work, that I here quote it,* often stating the history of the successes at agricultural shows of the certain cows with a few crosses from dairy cows. Mr. Carr says:—

"Mr. R. Booth's opinion that four crosses of really first-rate bulls, of sterling blood, upon a good market cow, of the ordinary short-horn breed, should suffice for the production of an animal with all the characteristics of the high-caste short-horn. In such an opinion, confirmed by such an example as this, there is much instruction and encouragement for tenant farmer's desirous of improving their stock. Female short-horns of high pedigree, are, in general, beyond the reach of their class; but if near neighbours would but club together, and procure for their joint use a succession of pure bred males, of fixed and determinate character, the improvement, in a few years, effected in their stock, especially as regards early maturity and tendency to carry flesh, would be such as materially to enchance their farming profits. It is often difficult, however, to convince even those amongst them who have used, and experienced the benefits resulting from the use of a high bred sire, of the expediency of continuing in the same course. Some wretched cross bred cow put to the 'pedigree' bull, probably produces a bull-calf with all the characteristics of its sire, all the more probably, perhaps, from her being of no distinctive character herself. This the farmer rears on something better than blue milk, in the hope of getting a

---

* The History of the Rise and Progress of the Killerby, Studley, and Warlaby herds of short-horns, by William Carr: London, 1867. Ridgway.

prize or two with him at local shows, nine out of ten of which absurdly ignore the first desideratum in a sire, pure descent, the bull of one cross being allowed to compete with the possessor of half-a-dozen. The mongrel gets the white ribbon, and immediately becomes, in his owner's estimation, endowed with every necessary qualification for a sire. The farmer thenceforth uses him to his own cattle, and perhaps those of half the neighbourhood. The result of this retrograde step is soon apparent in the stock. Interesting traits of the maternal ancestry of the *parvenu* bull re-appear in his progeny. The brindle, it may be, of Pat O'Flanaghan's Kerry, the black nose and horns of Sandy Macpherson's Kyloe, or the long-legs and flat-sides of Taffy Owen's Glamorgan. But though the farmer sees that he is rapidly losing all the ground he had gained, and that his stock has ceased to be sought after, he rarely admits the cause."

Now why should not the same fatal results follow from breeding bulls from the Irish Colonel O'Callaghan's red Scotch Polled Galloway cow? And at what stage in the descent do the produce became improved short-horns? The Galloway cattle at that period were noted as bad grazers, with very hard thick skins, and the red variety was always considered the worst. So much so, that it has, I believe, entirely disappeared, and black, with improved quality, has become universal.

That the bad handling and want of quality has continued in the descendants of this alloy seems admitted by Mr. Carr. However, admitting the want of what was considered the value of the improved short-horns, says the whole is a mistake and error. In speaking of a cow of this alloy blood which had distinguished herself at shows, and bred some prize cattle, he says:—

"The Isabellas had all the great capacity for rapidly acquiring ripe condition on pasture. As an illustration of the fallaciousness of the usual mode of judging cattle by the softness of their flesh, it may be worthy of mention that at one of the Yorkshire Agricultural Meetings held at Northallerton, a grass fed heifer, a daughter of Isabella, by Ambo, was shown and rejected as being too hard-fleshed. Not breeding, she was slaughtered at York for Christmas beef. Her two successful rivals also failing to breed were slaughtered, and the palm for the best carcase of beef was awarded to Mr. Booth's heifer over her Northallerton rivals. Nor is this the case without many a parallel in the history of royal shows. Numerous as have been the prizes which the Booth cattle have received, their number would have been greatly increased if judges had always carefully distinguished between flesh and fat. When their decisions have been on this ground—as they have often been—adverse to the Booth cattle, many an experienced butcher has proclaimed a very different opinion; and could the appeal *ad cramenam* have been adopted by an immediate sale of the rival animals to the shambles, how useless would it have been in most instances to contest the supremacy of the Booths!"

I fear I have exhausted the patience of my readers with opinions on handling and quality, and also that it is not loose fat. What could Charles Colling mean when he on his retirement said, "If I had my eye-sight and *the use of my fingers* I should not despair of another herd."

I cannot help asking too what farmers and dealers mean when, before buying lean cattle, they are careful to feel various points of the animals, and that from this test they regulate their biddings. Why are even the Irish dealers so eloquent in their invitations to customers. "Jist touch them yer honour, and for sure yer honour's fingers niver felt any to bate 'em! See the hair of them yer honour," &c., &c. I never knew a breeder of any knowledge use a bull without handling him, and at the Darlington bull markets, any animal with good looks and hair is handled and pressed for some purpose or other.

Butchers in purchasing fat cattle, certainly think it their duty and business to feel the animals before bidding for them. I believe handling and quality are not overlooked in any other breed. It will be an amusing result, if there has been this vulgar error in selecting improved short-horns as the most important breed of cattle in the civilized world. I am often reminded of the truth of the statement of Mr. Bates, that this alloy blood produces no prize bulls, by the fact that this alloy blood produced female prize winners before the bulls were noticed. Mr. Bates was doubtless correct when he said the judges at some great shows gave their awards without ever touching the animals.

As to in-and-in breeding, I believe Mr. Bates considered that it required the greatest judgment and experience. He had the difficulty of obtaining bulls of as good or superior blood to his cows; and whenever he departed from this rule he soon saw his errors. He had hired St. John (572) of Mr. Mason, and had thus the only female Baroness from the Duchess cow. The loss of Baroness was no doubt a serious blow to him. St. John had good blood in his pedigree.

Mr. Mason had used the best Colling bulls, and his herd when disposed of in 1829, infused good blood into many herds. The herd consisted of 64 females and 27 males. Lord Althorp purchased 16 females and 1 male. Mars (413), a son of Jupiter, ought to have been inserted in the statement as one of the bulls that Mr. Bates denounced.

Lord Althorp, as we see above, admitted that the St. John stock had weak constitutions. Mr. Bates calls it hereditary disease, but Mr. Fawcett informs me Mr. Bates spoke of it as consumption. Undoubtedly many short-horn cattle by in-and-in breeding, and high feeding and training, did become diseased. I often wonder that many more had not become so. We know well that diseased and rotten cattle existed, as appears from the following observations:—"It has been said that Robert Colling's stock were delicate. There is no foundation for this, and it may have arisen from the delicacy of Mr. Champion's cattle. Mr. Paley said the rottenness of the Warrior (673) family came from Diana, lot 3, and Mr. Champion has attributed it to Mr. Mason's Charles (127). Mr. Bates also attributed delicacy to Mason's St. John (572).* Mr. Dixon (page 126), in speaking of Mr. J. Grey's herd and his success as a breeder, says it was principally built up from General Simpson's North Star (full brother to Comet), and that he also bred direct from the Collings through Mr. Donkin's (of Sandhoe) blood. Young Star was the best bull he ever sent to Milfield, and Mr. Curwen and Mr. Blamire could not resist riding over to see him. The General had bred from Mason as well as Colling, but Mr. Grey did not care about the former, as he thought him tricky and all for form, and that his herd became hard to the touch and lacked constitution."

* Thornton, page 516.

PART IX.

Mr. Bates as an Instructive Friend—Mr. Rhodes on his own Stock and the appreciation of Mr. Bates—Mr. Bates on Pedigree—Mr. Coates, Mr. Whittaker, and the Original Herd Book in MS.—Remarks by Mr. Coates on the Duchesses, and the late Mr. H. H. Dixon's partiality—Lord Althorp on the Herd Book—The Almighty Dollar—Mr. Bates and Mr. Grant Duff on Stock-pedigrees.

Mr. Bates often gave his advice and assistance to his friends, especially to young breeders in obtaining cattle, and I insert a letter to show his feelings towards the Rev. J. Armitage Rhodes,* then of Carlton, near Pontefract, a gentlemen with large estates in Yorkshire, for whom he had purchased a cow.† It was written to Mr. Whittaker, the well-known breeder of short-horns, and a mutual friend. His Grace had been let, not sold to Lord Althorp. The Duke (226), was the Colling bull, brother to the Duchess I. Lord Althorp had sent some cows to him. Mr. Bates had offered to bear any loss by the cow Spot he bought for Lord Althorp from an accident which befell her, and Mr. Bates expressed the same readiness in the following letter when his purchase produced a *black-nosed calf*:—

"Ridley Hall, February 19th, 1822.

"My dear Sir,—His Grace reached Kirklevington the evening before I left and had borne his journey well considering the effects of the severe cold he had had, and which I fear, from his thin

---

* This gentleman died in May, 1871, aged 87 years.

† Kate, Herd Book, volume I., page 353. Bred by Mr. Pilley, property of Rev. J. A. Rhodes, by Mr. Pilley's Wellington, dam by Mr. Pilley's Eclipse. Bought for 37 guineas and sold to Mr. J. Hunt for 35 guineas. (G. Coates.)

appearance, was not likely soon to get quit of. Ketton the Third also got safe up, and was so for some days afterwards, when he began swelling, and they had to tap him frequently while I was there. I am really sorry to hear our friend Mr. Rhodes has been so unfortunate with all his cows, and the calf having a black nose, and Lord Althorp saying Pilley's cattle are all so, has surprised me not a little. I did hope our friend had been well laid in as a beginner at Doncaster, but the other cows having all cast calf, and this calf with the black nose being also dead, hurts my feelings very much, particularly as I found Mr. Rhodes so very tractable and willing to follow advice. The other cows could all be got to pay for feeding without any loss, but this cost, I think, about the double of butcher's price when fat as markets are at present. I do assure you I thought well of her at that time, and had it not been to serve our friend would have bought her for myself, and notwithstanding all you say, and I have no doubt justly, as his lordship's report may well be depended upon, and notwithstanding that his lordship will have no more of Duke's stock, if our friend Mr. Rhodes chooses to part with the dam of the black-nosed calf I will yet give him prime cost, and if he likes to part with her would send for her and put her to The Earl, son of Duke, dam out of the own sister of Duke, and the first bull calf she brings shall be at his service, if he chooses to keep it as a bull *solely* for the use of his neighbours' cows in his own parish and his own, thus promoting the patriotic object he has in view and a truly laudable one it is. Do write and inform him of my offer, and keep up his spirits, and tell him if he is not quite disheartened, and will trust me again to choose for him, I will keep looking out for another which I hope will be more fortunate and never bring a black-nosed calf. To The Earl she will never breed a black-nosed calf, if I have any judgment whatever of cattle. I do really wish to have her to redeem both the cow's character and my own judgment, and I trust our friend will not let his delicacy overcome him if he wishes to part with her, and if he does I should wish to have her here immediately before she takes the bull, therefore will thank you to let him know my sentiments on the subject, and give him your advice what you think he ought to do without any reserve whatever. You were very desirous I should sell him the heifer out of Duchess the 4th, by Ketton 3rd, when at Kirklevington last spring. She is now in calf to The Earl, bulled the 19th December last, and as fat as when you saw her in June, and if you thought the offer of her for our friend at fifty guineas, and the dam of his black-nosed calf, which cost about thirty-seven pounds, I will give him the choice. For this heifer I would not take one hundred guineas; it is therefore giving above fifty for his cow, notwithstanding the black-nosed calf she has brought, and I would be at the expense of sending this heifer to him and bringing his cow here; advise him as you think best for his interest.

"I have now two bulls by Duke, out of the dam of Ketton 3rd, and a heifer by Marske, and out of the same cow, and bulled by The Earl, and for the three I would not take 3,000 guineas, bad as times are for farmers. I make the offers through you, that you may give him your advice, but desire him also to follow his own after all.

"Lord Althorp blames Duke, when the disorder in the calf was from Rosette's tribe, a discovery I made from having Red Rose's own brother, Styford, hired in 1804, of Mr. R. Colling, and you may remember a cow, the third generation therefrom, at Kirklevington, when Mr. Ellis was with you from Redcar, after Major Rudd's sale in 1819. I had not one of that bull's (Styford) get, which did well, and I put forty cows of my own, and near twenty of my father's, and they also were the same. Old Ketton's stock were the up-making of me, and now that I have again got the blood, pure of other mixtures, shall never again part with it for any other tribe of short-horns I have ever seen. You would, I am sure, spend a pleasant time at Wiseton with Lord Althorp. He is a truly humble gentlemanly man.

"I did purpose on taking my pen to have given you my pedigrees for Coates. If I knew when he had all ready prepared, I would give him a meeting, either at your house or elsewhere, and give him every information in my power, and, previously to his publishing, it would be right to show the whole to Mr. C. Colling, and if you would come there (to Croft), I will meet you at any time, by having timely notice. *To have the whole read to Mr. Colling, two persons being present, or perhaps our friend Mr. Rhodes would make a third, and hearing Mr. Colling's remarks, would prevent any after disputes about pedigrees. If this is not done, they will be endless.* Hutchinson is sure to attack this book, and, notwithstanding all I have said to him, he has published quite the contrary to what I had told him I knew and had seen myself.

"I am truly sorry that you have been so unfortunate with your cows; it is very disheartening. I lost in four years, 100 head, but the last seventeen years, since I bought the first Duchess, my stock have done well, and though this place had the character of stock being often ill, I have not had anything unwell as yet, no farrier or cow doctor has ever been here."

In reply Mr. Rhodes wrote as follows to Mr. Whittaker:—

"Mr. Bates appears to me to take up the subject of the cow under an erroneous impression, as if I were disappointed and disgusted with her. The fact is that the circumstance of the black nose to the calf, may in a great measure be anticipated on account of the known defect in its size. The frame of the cow is, as far as my imperfect knowledge can enable me to decide, very superior, and if so, the price I gave was not worthy of consideration. But if I had in truth been 'hung on' with her, as it is vulgarly termed, I trust I should be one of the last men to wish to be relieved from any pecuniary loss by 'taking in' any body else with her. Except the circumstance of being torn behind, she has no defect that I know of, and she is likely to be the mother of excellence milkers, as her bag is large, and she milks very well indeed. She has become so quiet that I can lead her anywhere. These circumstances, naturally, have acquired her a sort of individual favour and attachment with us, and we are therefore quite indisposed to part with her on any idea of imperfection or insufficiency. But, at the same time, I have received so many marks of favour and regard from you and Mr. Bates, that I am willing to accede to his wishes, and to be guided by your judgment. If you think that I have a wrong opinion of his notions, and that he really estimates the cow in the same way that I do, I will yield to your suggestions, and put the matter into your hands, on only this promise that Mr. Bates, and not you, shall make the compensation you think acquitable, that no advantage shall be taken of Mr. Bates's constitutional ardour, and that the payment shall be in cows and not in money."

In February, 1822, Mr. Rhodes wrote Mr. Bates:—

"Mr. Whittaker has this day forwarded me your letter respecting Mr. Pilley's red cow and I beg you to accept my thanks for the offer you have made, so that I should be no loser by following the advice with which you favoured me at Doncaster. But, my dear Sir, I should be ashamed of myself if such a consideration could ever find admission into my mind. I have the fullest confidence that the opinion you then gave me was perfectly justified by appearances; and the misfortune which has since happened of a black-nosed calf, does not lead me for a moment to question the accuracy of your judgment. I beg you therefore to believe me very much your debtor for this as well as many other acts of kindness. If this cow upon further trial should still show an alliance to the black-a-moor, I must try again to find something more likely to answer the purpose which I had originally in view, and any small pecuniary loss can be of no consequence whatever. Had I been admitted to more extended acquaintance with you, I flatter myself that you would not have given yourself a moment's concern on the subject. I have ordered a horse-hoe to be made, of which I beg your acceptance as a small token of my esteem and gratitude. A grubber has been also made for you, according to your order when here, and as I think the implement extremely valuable, it will, if you think fit, be sent with the horse-hoe. The price will be about £4 10s. I think. I did not countermand it because I believe it will be found most effective; but it is not at all necessary to take it unless you like as they are frequently wanted here, and quite supersede the plough in many cases by the use of them. The saving is very considerable and the effect in the land highly advantageous. We were much gratified by a little tour into Nottinghamshire, and did not meet with any thing like Major Rudd's uncourteous conduct.

"Mrs. Rhodes joins me in kind regards. It will give us pleasure to hear that you are quite restored to health, and to full enjoyment of all those blessings which Providence in its bounty permits you to partake."

"You did me the honour to address to me a letter on the 31st

October. I am happy to hear you have recovered your health, and that your farming concerns are going on satisfactorily, although your prognostication as to the rise in the price of wheat is rather tardy in being realized. However, the dock roots you say were abundant in the newly acquired stubbles, and you may therefore hope that with careful management, you may have a nice young family to take under your care next year. I have this morning a most kind and affectionate letter from Mr. Coke, of Norfolk, in which he invites me and my wife 'to spend several weeks with him at Holkham.' His opinion of the state of agriculture is of a gloomy kind, and he cannot apparently raise his mind upon those buoyant spirits which always support you. In order to console me for the loss of the Herefordshire tour, we have been making a little trip into the County of Nottinghamshire, to see Lord Althorp's, Mr. Champion's, and Mr. Simpson's stock. We were much gratified, and he has added many new particulars to his history of short-horns. I have recommended him to publish, promising to use my influence with Mr. Hutchinson to write him a poetical preface, in the same classical taste as the specimen, which was lately inserted, with other accurate and delightful observations, in the *Farmers' Journal*. Warren, 'hide thy diminished head.' Day and Martin[*] creep into insignificance and obscurity, for who will hereafter read the praises of Japan blacking when sockbreeders shine. Orlando, Firciso, and Norman Willie, shall dance amidst all the flowers of poetry through the wilderness of fancy and romance. I have been particularly unhappy in my first attempts to acquire celebrity as a breeder of short-horns. I bought Marchioness and Cecilia of Mr. Whittaker, at Ferry Bridge. I sent Marchioness to the house of a friend, for fear of infection from the cows which had here miscarried. On the third she presented me with a beautiful dead calf, about half-grown. Cecilia was taken ill without any previous indisposition on Monday noon last, and before midnight she was dead. Her complaint was the quarter-ill in a most virulent degree. Two of my purchases at Doncaster kick in the most persevering manner, and must be parted with forthwith; but the red

[*] The celebrated Blacking Manufacturers, who kept 'Poets' to puff off their goods.

cow is well, quiet, and likely in all appearance to produce a calf in the proper course of nature. The preceding misfortunes have given me pain, but, as it was impossible to prevent them, it is fruitless to repine. A bountiful Providence has given me many other sources of happiness, and to them I must gratefully turn. My wife joins me in respectful and kind wishes for your health and well-being. We regret you were not able to make the drill act well, but we shall still send you, with your permission, a horse hoe, for your turnips, peas, beans, &c., in testimony of my sense of obligations for your many kindnesses.

"P.S.—Young Daisy is well, and we think much improved. We believe he has been the cause of the miscarriages by a trick he has of butting the cows. We have therefore got him a pair of spectacles to peep through."

Mr. Bates, acting on the notion that pedigree was everything, and being satisfied that the Duchess tribe would continue to flourish, adopted the plan the most likely to distinguish the breeding of the tribe, viz., of numbering them like a royal succession. In no other way could his judgment have been so completely seen as by the numbers he adopted. It is not surprising that he should wish for some authentic record of the breeding of the improved short-horns. No other breeder did, or perhaps could, adopt the same plan. Mr. Mason entered his cows in the Herd Book, numbered from 1 to 60, but not in families. They had no succession of stock which could be so distinguished, and after the death of Baroness, in 1812, he called his purchase of 1810 Duchess 1st.* This, together with the alloy, or grandson of Boling-

* Notwithstanding the care I have bestowed on the subject, I feared I had fallen into an error about the price of the cow Duchess of 1804, but Mr. Bates is correct when he says he gave the first 100 guineas the Collings got for a cow, and that this was in 1800. Of

broke controversy, no doubt set the herd book of George Coates on foot. The above letter shows the anxiety of Mr. Bates in 1822 that it should be correct and authentic, and he frequently expressed his regret that no restrictions were placed on the insertion of pedigrees by any one who paid for it. George Coates had been a tenant of, and lived near Mr. Rhodes. After G. Coates had collected all the pedigrees that were to be inserted, the whole were written out by Mr. Jonas Whittaker, in a very neat, clear hand, and the remarks and observations of George Coates were inserted in red ink in the same hand. This book has generally been considered the manuscript, but it is only the matter, and was altered in form and shape, and the red ink parts omitted. It was presented by Mr. J. Whittaker to Mr. Rhodes, and given by him to his nephew, Francis Darwin, Esq., Croskeld, Otley, who, at the request of his friend, Thos. Bates,* Esq., of Heddon-on-the-Wall, Northumberland, has sent it for my use, and it is now before me. Mr. Darwin was a contemporary of

this cow I have no trace. In 1804, he gave also 100 guineas for the cow, and it was for her calf, the dam of Duchess 1st., that he gave not one-third of her price, 183 guineas. Sir H. V. Tempest in 1801 gave much more.

* This gentleman succeeded to, and is now, the owner of his uncle's estate at Kirklevington, on the death of his elder brothers, George and John Moore Bates. He was formerly fellow of Jesus College, Cambridge. My best thanks are due to him for the use of all his uncle's letters and papers, and for the continuance of the friendship and confidence I had enjoyed with his late uncle. I deeply regret that his other occupations did not permit him to devote his time to the continuation of the herd in the family. He has possession of Passion Flower the 4th's daughter. Although not entered in the Preston Hall sale list, was bought there for 20 guineas, and is now in calf to 13th Grand Duke.

the Rev. Dr. Bates at Christ's College, Cambridge. I have perused this book with great interest to see how far it supported the statements of Mr. Bates, and I find the red ink notes are in full accordance with Mr. Bates's opinions. There are also some marked T. B. or Bates as the informant, but not those I now quote. I hope to add in an Appendix the pedigrees of the most important animals which appear in the Kirklevington pedigrees, or are mentioned by me with the above notes. I have, however, given those objected to by Mr. Bates, viz. :—

BROKEN HORN.—Red roan, 1787, bred by Robert Colling, by Hubback, d. by Hubback, g. d. bred by Mr. Watson, of Mansfield, brother to the g. g. d. of Robert Colling's Red Rose.

PUNCH.—Red yellow, bred by Robert Colling, by Broken Horn, d. g. Broken Horn, g. d. cow of G. Best, of Mansfield, good size and form.

BEN.—Red, bred by Robert Colling, by Punch, d. by Foljamb, g. d. by Hubback, own brother to Robt. Colling's Red Rose's dam, great hind-quarters, but shoulders not good.

FAVOURITE was born in 1793.

Mr. Bates said there were never two crosses until Favourite was used. I can find no trace of such crosses. Hubback was only in use two or three years, and his produce must have been very young when put to him. Some one had evidently taken great pains with the above pedigrees to get the full benefit of the Hubback blood.

Mr. Dixon had seen the above manuscript book, and quotes the opinion of George Coates, that Duchess 1st was only "fair," but does not give the other opinions, which are :

Duchess 2nd	Sisters	Crops. Hind quarters fair.
Duchess 3rd		Fair.
Duchess 4th	Sisters.	
Duchess 5th		
Duchess 6th	Very Good.	
Duchess 7th	Good.	
Duchess 8th	Sisters. Both Good.	
Duchess 9th		

All writers agree with Mr. Bates that Duchess 1st was inferior to the Duchess of 1804, her grandam, and that Colling's bulls, even Comet, had not improved the Duchess tribe. The above corroborates the assertion of Mr. Bates, that he had by the use of Ketton No. 1, improved on the Duchess No. 1, of Charles Colling.

Mr. J. Whittaker wrote respecting the Herd Book to Mr. Bates as follows:—

"Burley, 16th Feb., 1823.

"My dear Sir,—Coates' book being now finished, except a few prints which are expected from London this week. I beg to ask you if you would have yours forwarded. There are several subscribers in your neighbourhood, and if Mr. Coates could send them altogether it would be less expensive to you; and if one person could receive and remit the money I would be glad as I have advanced the poor old man a little money on account, and I would feel much obliged if you could procure him a few more subscribers. The printer having omitted Ketton 1st and 2nd, you will find them in the appendix, which cannot make any difference, as the index refers to them. Besides the prints expected this week from Mr. Arbuthnot, there will be several more, one of which will be Comet, which will be forwarded hereafter. I should be happy to hear it has your approbation and sanction, and that the old man may have something to support him in his old age. *The book is merely a copy of the pedigrees furnished by the subscribers, so that, although errors may appear, he cannot be blamed.* I have attentively perused it and think favourably of it.

"Mr. Jetty has a bull calf by His Grace, which he thinks too good for a steer, but the agreement betwixt you and him prevents him from keeping it a bull without payment of a sum of money, and which the times will not warrant him in doing; but he will not cut it until he hears from me."

Mr. Bates was not the only person who detected errors in the Herd Book. I quote Mr. Dixon's account of Mr. Grey's visit to Lord Althorp:—

"When Mr. Grey called upon him at Downing Street, and saw 'George' as a preliminary. The latter remembered him and gave a little dry laugh: 'You've come about cows, Sir, so you'll not have to wait long.' Sure enough his Herd Book lay beside him on the desk when Mr. Grey was announced, and formed the text for the next half hour. Every Monday morning his lordship received the most accurate budget of what cows had calved during the week, with the calf marks, and he did very little work till it was all transcribed into his private herd book. This morning he handed Mr. Grey a letter. 'There's a letter,' he said, 'from Carnegie;' he admires my political course, and he writes from the Lothians to say that I shall have the first refusal of his bull.' Then he so characteristically added—'I've written to thank him for his political confidence, but I've told him there is a flaw in his bull's pedigree. He traces him back to Red Rose, but Red Rose never had a heifer calf.'"

Mr. Bates showed great kindness and hospitality to one American gentleman, who ingratiated himself so much with Mr. Bates, that he presented him with a bull, which was taken to America, and the half share of him sold there for 150 guineas. Mr. Bates was then asked to give a receipt to show that 300 guineas had been paid for him. Mr. Bates, however, took no notice of the request, and gave orders that if ever the person came again to his house he was to be refused admittance, and that he would not see him on any account. Mr. Bates paid no regard to the reported high prices paid for short-horns by private sale, as he knew so many instances of receipts being given for sums far beyond the real prices.

Mr. Bates's indignation was often roused by the methods sometimes used to make large sales, and to create a competition to force up high prices. One sale caused him great pain. A number of cattle had been bought from farmers in the district, and then were entered with long pedigrees in the sale of a large and well-known breeder of short-horns. I was present at the sale, and stood among the farmers and heard their observations, such as follows, when an animal was brought out:—"Why that was mine. I could not give him a pedigree. How has he got it?" And at another sale, Mr. Bates went early to inspect the herd. I give his own printed account:—

It is not three years ago, I went to see a stock advertised for sale, and they were all stated in the catalogue of the sale to be descended from one animal, whose represented sire I had used, and he was cut when he left me, and I saw him soon after being castrated. The parties whose stock was for sale were honest upright characters, and told me the year their father bought the calf; I told them I had had the use of that bull and had seen him cut, after having had the use of him one year, but that was more than six years before this calf was calved: they assured me they were totally ignorant thereof, but that the person they had appointed as auctioneer assured them such was the case. On the day of sale, on the auctioneer arriving, he called me out to a private place, and told me he was as certain of the fact as he was of his own existence of the bull they were descended from, and I have little doubt he would have taken his oath it was true, although he had proof given him by an eye-witness of the fact, that the bull was castrated above six years before the calf was calved; and then turning round, and pointing to the large company that was assembled, he said, "you see what a company I have brought together, and what a high sale I will make by my pedigrees of this sale." And the same auctioneer has continued the same pedigree to the same animals since he was certain it was a fabrication. I mention this, by the way, to guard persons against the prevailing pratice of pedigree making.

The sale took place according to their pedigree, and many herds now boast of animals descended accordingly.

Mr. Bates did not insert any pedigrees in or subscribe to the Herd Book after the death of Coates, and he endeavoured to obtain accurate pedigrees by some official or authentic parties, and to establish a rule that no animal should be inserted unless likely to be of some importance in future pedigrees. He objected to insert animals and keep a record of mere fat and dairy stock. He communicated with several influential gentlemen on the subject, and it was thought the best course to have yearly reports of all the births (which I believe is the one adopted with blood horses), and from such lists those thought worth preserving must be selected and the lists published by authority. Mr. Bates considered that such lists, if submitted to the Local Societies or some such authorities would secure their accuracy, and so stop any attempt at making a pedigree when an animal turned out a good show one. Mr. Bates received many communications on the subject. I insert one from Mr. Grant Duff:—

"My dear Sir,—I think you know that in order to prevent deceptions, or falsification of pedigrees in the North of Scotland, I introduced annual lists which have, I believe, been of some little use. And, although I may soon retire from the stage as an extensive agriculturist, and my herd of short-horns now only number twenty, I have continued to issue the lists, and the space on the sheets enabling me to extend remarks. You will perceive I have taken the liberty of quoting your opinions in regard to the injudicious manner of giving prizes for beef instead of blood. I hope you still enjoy as good health as when I had the pleasure of seeing you at York.

"I have always publicly and privately expressed my regret at your not giving the world your opinions of breeding and general management of cattle. No doubt rivalry and the disagreeable necessity of replying to opponents entail obstacles which a wise man is unwilling to encounter, but if you were privately to distribute copies amongst your friends, as I do these lists, the good might be spread without incurring the evil.

"Edin, by Banff, Nov. 9th, 1848."

The principle of Mr. Bates that, in exhibitions of breeding stock, the pedigree ought to be fully considered, has now been admitted, and the importance of it has been recognised by the Royal Agricultural Society of England, who have adopted regulations to the effect that "Each animal entered in the short-horn classes shall be certified by the exhibitor to have not less than four crosses of short-horn blood which are registered in the Herd Book."

Mr. Bates would have been precluded from exhibiting by this rule, as he was resolved to enter nothing until the lists were verified, and he objected to a private register in which any person might enter anything they paid for, and to the price of a publication of which there was no public control. It was a subject he said of more than national importance that the breeding of short-horns should be free from doubt or suspicion.

PART X.

Cattle never Doctored—Libel on 2nd Hubback Refuted—by Epidemic—Intercourse between Breeder and Butchers—A. Maynard: his Stock and Character—Kenyon College—American Skill and Wisdom—The Value of the Duchess Tribe—Sales at America and Tricks in the Trade—American Cow—Appreciation of the Kirklevington Cattle in America—Shows and Prizes.

Mr. Bates was very candid and liberal with several persons, especially young breeders and buyers, who obtained stock from him, or who acted under his advice, but he met many ungrateful returns for his kindness, the parties being unable to understand his motives, and bent only on obtaining all the advantage they could from him. I may here remark that Mr. Bates, after the first few years at Halton Castle, never employed a cattle doctor, or bled, or blistered, or gave any medicine to his stock. He sometimes lost young stock with quarter-ill or blank-quarter. A recent publication, in noticing 2nd Hubback, says, that Mr. Bates, in relying "on his idolatry for this bull, did his herd no little harm; and it was only when he found that he had lost 28 calves in a year, solely through lack of constitution, that he had to cast about for another bull."* I can give this account the most distinct contradiction. Mr. Bates lost a great many calves in two years together at Kirklevington, but it was bronchitis or hoodle, all the air passages in the lungs, and up the wind pipe to the gullet were full of small, thread-like worms, which caused them to cough incessantly. I believe the complaint originated from drinking stagnant water during the

* Druid in "Saddle and Sirloin," page 154.

heat of summer. In Clator's Cattle Doctor it is stated that, in the spring of 1831, thousands of young cattle died in every part of the country of this complaint, and that it often appears in the form of an epidemic.

For some years before 1833 very few short-horn breeders visited the herd of Mr. Bates, and he took no steps to bring it before the public. He reared and fed his steers, and also fed the cows and heifers when they did not prove in calf within a certain period, and many a noble female was thus sacrificed. He never had calves born during the three summer months.

When in Northumberland, Mr. Bates very seldom sent any fat cattle to the market. The principal butchers in Newcastle and Shields came to buy his stock at home. They were a most respectable and well-educated class, and in their business transactions they turned over very large sums of money weekly, and they generally stood in a high position amongst the traders in their towns. Mr. Radcliffe, mentioned by Mr. Bates, was a frequent visitor and purchaser, and also Mr. Walker, of South Shields, who carried on a very large and lucrative business, and whose grandson was the first M.P. for that borough. Buyers often inspected the whole herd and gave their opinions on their defects and merits. The breeder and feeder was thus made aware of the defects and merits when they were to be tested in the scales; and for well-bred and properly-fed animals they always obtained a higher price per stone than for coarse and ill-shaped ones. The breeder and the buyer were thus enabled to exchange their opinions and to profit by their intercourse in a manner impossible when cattle were sold by auction or by agents without the presence of the owners.

From about 1820 to 1833 Mr. Bates rarely let any bulls, and kept no calves for bulls except what he thought he might require for his own use, or induce his friends to introduce into their own herds for improvement of the stock. Mr. Rhodes used Enchanter (244); and in July, 1823, Mr. Rhodes wrote as follows:—"The Red Cow I bought of Mr. Pilley has brought me at last a bull calf, by Enchanter. It is rather lighter coloured than she is, and the nose correct. If you wish to have it you have only to say so."

Mr. Anthony Maynard, of Marton-le-Moor, a well-known agriculturist and sportsman, was always an admirer of the stock of Mr. Bates, and used Laird bull. Mr. Jonas Whittaker, in November, 1824, writing to Mr. Bates, says:—"What was Laird got by? I saw his descendants at young Mr. Maynard's at Marton, and they were most extraordinary animals. Can you inform me of the bulls' names used at Acklam, beginning with the first and downward in succession. I see the Grey Bull is said in one place to be by a son of Favourite, and in another by the son of the White Bull. If I ask the breeders there they refer me to Mr. Bates. Mr. Rawdon Briggs is about to abandon short-horns. His stock is wretchedly bad. What do you say of Mr. Berry's production? I hope you saw Mr. Charles Colling. I think it must be pleasing to him, as he is the chief of the piece. Mr. Wm. Wetherell says the heifers I bought of him were by Col. Chaloner's Major by Miner, dam by Petrack, &c. Will you be so good as call on the bookseller, and if he has received for the remaining books of Coates's, you will please to hold it on my account."

Mr. A. Maynard was brother of Mr. J. C. Maynard, of Harlsey, the purchaser of Marske, and was also the owner of the bull Match'em and the cow Portia and a well-known herd. I venture to insert the following as a tribute to his memory *:—

"If there is ever a gallery devoted to the heroes of 'field and fold,' the late Mr. Anthony Maynard will infallibly have a place. He came from quite a short-horn and horse-loving family. 'Maynard's bull' is a name of note in the 'Herd Book,' and 't'auld yellow cow," to which he so often reverted, made her peculiar mark. Crusade (7898), by Cotherstone, by Bates's Cleveland Lad, from a granddaughter of John Colling's celebrated cow Rachel, was his most famous show beast; but he had done nothing in that way for some time past. He leant decidedly to the Bates blood, but bullocks were his secret pride. He delighted to recount what toppers (the best of which was nearly lost in the snow) he and his father before him had pitched at Yarm; and how both of them would take 'to boot and horse lad' and ride thirty miles across country by daylight to be at market betimes. He was always a very active man, a keen sportsman, and rode well to hounds; and it was, we believe, a hard-riding accident which caused that peculiar crick about one shoulder which, with his keen, intelligent face, made him 'so good to know.' For twenty years he kept the Boro' Bridge harriers, and showed excellent sport. The Raby country then extended as far as Boro' Bridge, and the Duke always charged him, '*If you find an outlying fox do your best to handle him before he reaches a cover.*' He hunted both with the Bedale and the Raby, and when either of the masters appealed to him at a check: '*Which way, Anthony?*' the general reply was '*overridden by those young officers—cast behind them.*' On hunting days he was up at five, and rode over his six hundred acres before breakfast, and then fifteen or sixteen miles to cover; and no man told better Yorkshire hunting tales over a bottle of '20 port. He was one of the oldest short-horn breeders in the kingdom, and we hear that his herd numbered about 120 head at his death. To

---

* Druid in "Saddle and Sirloin," page 192.

the 'Herd Book' he had been a contributor since its commencement, and his numerous entries traced to good and ancient families.

Marton-le-Moor, a few miles from Ripon, was his pleasant, old bachelor home. The handsome Crusade, with a portrait of his owner and his herdsman, formed a leading feature of the snuggery, and a large painting of 'the best side of Comet' (as he did not fail to tell you), held the place of honour in the dining-room. A Yorkshire show-ring hardly looked itself without 'old Anthony' or Crofton inside it, and he was quite regarded as a 'chief justice' in short-horn matters. A more upright judge did not exist, but he had very strong dislikes and 'crotchets,' and did not scruple to express them when he was not on the bench. To the Butterfly tribe he was never reconciled. The Royal had his services as judge at Chester, in 1858, and again at Leeds, in 1861; and he liked the business so much, that, when he was verging on seventy, he crossed the Channel to officiate at the Dublin Spring, and proved himself in the possession of wonderful 'sea legs.' In judging he generally gave more points for mellow handling than for gaiety of form. He went not so much for size as for neatness and quality, and at Dublin he was in the minority when Rosette was drawn against Sweetheart for the first cow prize, and he took good care to let his opinions be known. He couldn't see it at all, and led many a breeder up to the pair in the course of the day, and with that odd jerk of his stick, proceeded to argue out the point on which 'Mr. Stratton had been so stiff.' Very few, if any, had finer taste, but he was not free from that peculiar cynicism in describing a beast which is the vice of so many good judges. Part is spoken of as though it were the whole, and there is no balance of points. Thus, if he spoke of Belleville, he would say, 'If you backed his hind-quarters into a hedge he was good enough,' and left it; and unless you pressed him hard you heard nothing of his beautiful head and fore-quarters, and 'soft, molelike skin.' We believe that he had been at very few shows since the Leeds Royal, and that for many months back he has been in a very failing state—so much so, that it was hardly thought that he would see the New Year in. He was one of the last of those 'grave and potent seniors' whose fine experience we can so ill afford to lose.',

Mr. Bates had many years ago been introduced to the Bishops and gentlemen from America who visited England to make arrangements for establishing a college in connection with the Episcopal Church at Gambier, on the Ohio. This was called Kenyon College, after Lord Kenyon, who, with many English Bishops and noblemen, took a great interest in the work. Mr. Bates was very active in his support of the establishment. He was much pleased with the intelligence of the American gentlemen, and many of them visited him in Northumberland and Yorkshire, and strong inducements were held out to him to emigrate to America and settle on the Ohio, in the neighbourhood of Gambier; and he seriously entertained the project, and contemplated selling his estates and removing thence with his whole herd. The conduct, however, of one of the party who had been some time resident with Mr. Bates became so unsatisfactory that he abandoned the project.

The talent and enterprise displayed by the inhabitants of the Western World has in nothing, perhaps, been more successfully and usefully developed than in the establishment of herds of short-horn cattle. I had written many pages to show the capability of the United States and Canada for the maintenance of large herds, and the importance to the country of producing hides, tallow, and beef on the most economic principles; how they could, in fact, manufacture almost innumerable products of their soil into marketable products. After all, cattle are only the machines by which this is accomplished. In the United States and Canada short-horn cattle are not the fancy or hobby of a few gentlemen or noblemen of large independent fortune. They are the investments of highly-educated and experienced sensible men of business and commerce. Even the Society of Friends and the Shakers join in giving heretofore unheard-of sums for cattle of the Kirklevington blood. Who can say that the "preference for Mr. Bates's blood, especially that of the Duchess tribe, is a mere caprice on their part, and that it is not founded on reason." In announcing the departure and arrival of Duchess cows and heifers to and from America, a writer says:—"John Bull sends his pretty daughters to the States. Jonathan ungallantly sends them back again and 'calculates' the profits. If stars and stripes are not badges of nobility, a pair of horns appears to be, for the red, white, and roan is registered with more titles than the Premier peer."

I did intend to enter very fully into the history of the Kirklevington cattle in America. As, however, I understand that Mr. Lewis F. Allen, the editor of the American Herd Book, is now engaged in writing the history of the American short-horns, I shall only mention a few to show the progress they had made in the lifetime of Mr. Bates. I cannot describe the mingled feelings of pain and pleasure that I feel when I read the now almost weekly announcements of the departures and arrivals from America of the descendants of the Kirklevington herd. The sensation created by the sale at Windsor of the Geneva Herd from America, mentioned hereafter, is now superseded by the daily announcement of the recent prices, which prove most conclusively that the estimate of Mr. Bates of the value of pure blood were far below the proper standard. A Northumberland gentleman still alive relates his astonishment when he was told by Mr. Bates that the descendants in his herd of the Duchess cow were worth £10,000. I often call to mind the repeated statements of Mr. Bates that the Americans knew the properties and value of improved short-horns much better than his own countrymen, with very few exceptions, and that the proceedings of the Royal Agricultural Society in awarding prizes as they did to short-horns, would have the effect of sending all the best to America, and that our own countrymen would have to go to America for their breeding stock.

Mr. Bates also dilated very much on the superior advantages enjoyed by the Americans in the maintenance of their herds. The boundless plains and prairies were constantly mentioned by him and pointed out as the proper scene for enterprising and experienced breeders. Entire freedom from rents, rates, tythes, and the full enjoyment to the occupier of all improvements in the land. Also an intelligent and highly-educated population, free from the narrow prejudices and interested motives which made so many persons attempt to depreciate his herd.

Mr. Bates frequently expressed his regret that he had not in his youth had such a country as America for his energies.

On Easter Monday, 1833, I was as usual with Mr. Bates, at the Darlington Great Market. Some American gentlemen were there, at the King's Head Inn. They represented the Ohio Agricultural Society, and had come to England to buy short-horn cattle. When they came to Liverpool, they enquired where the best cattle of this breed were to be found, and were directed to Darlington; when they enquired after short-horns they were advised to apply to Mr. Bates, and did so. Mr. Bates soon got into conversation with them and found they had a great knowledge of the subject, and he invited them to Kirklevington, and showed them his own herd, and took them to view the principal herds in the neighbourhood. A cow, called the American cow, and sister of Styford and Red Rose had, many years before, been sent to America, and been brought back, and probably she was the first improved short-horn sent to that country. Mr. Bates was very anxious that the Americans should obtain the best breed of short-horns, and very much to the surprise of his friends, he offered them the choice of two out of six females, and he finally sold them Rose of Sharon, or Red Rose 11th, and Teeswater, the former at 150 guineas, Mr. Bates voluntarily engaging to furnish the first sister-in-blood, gratis, if she did not produce a living calf to the buyer. The latter was sold for fifty guineas, and in like manner, if she did not bring a living calf, the buyer to have a heifer, gratis, off the cow, two crosses by Waterloo and then Belvedere. He also sold them a bull calf called Earl of Darlington, and another bull calf called Young Waterloo. This was in April, 1834.*

Mr. Bates offered to procure them any cattle which they might require, provided he could obtain such as, in his opinion, were proper tribes, and he bought a bull

---

* Rose of Sharon bred only two calves—a bull and a heifer. Their descendants have been prize cattle, and known as the Cambridge Roses. I believe the others also proved most valuable and were all much admired in America.

from Mr. Clark, of Skipton Bridge, which he recommended them to take, but they declined. Mr. Bates introduced the American gentlemen to a breeder in the neighbourhood, who offered them a bull; which, however, they declined. In the following year Mr. Bates received a letter from America, asking him to procure several animals for the society; but, a few days after its receipt, the agent or manager of a gentleman whom he had introduced to the Americans, called at Kirklevington, and wished to buy cattle, saying that he had received a commission to buy for the Americans, and Mr. Bates did nothing more, his friend having thus got into a lucrative commission business; and this agent actually purchased and sent to America the neighbouring bull which had been rejected the previous year; and, after viewing Mr. Bates's herd, told him that Mr. Clark's bull was the best he had seen in his journeys to buy for America, and offered Mr. Bates 150 guineas for him; but Mr. Bates told him that the Americans had declined to take him the year before. The buyer said that was of no consequence, as they would not know him; but Mr. Bates said he would take care and let them know; and so the bull was refused, and Mr. Bates presented him to the Kenyon College, at Gambier, for the use of the college cows.

Mr. Bates warned the commission buyer against the blood of Hermit (305), who replied "The Americans know nothing about Hermit; what they want is a long pedigree." Mr. Bates then wrote to America thus:—

"I told him Hermit he knew and had seen and had ruined one of the best tribes of short-horns, and that stock so descended would bring the best short-horns into disrepute. If you had only looked at the predecessors of the bull, which you gave £175 for, in Mr. Whittaker's sale bill, which I gave you, you would have seen how low they stood in estimation, from the prices they were sold at Now, by having put this bull to the heifers you got of me you will bring their produce into disrepute; and I will, on no consideration whatever (if you would give me ten times the price I would otherwise have charged you for a heifer) sell you any to put to any bulls but what I have bred, or are of my blood. Nor will I sell you any at any price till you and the Company you act with, under your joint hands, have solemnly promised not to do so. My object has never been to make money of breeding, but to improve the breed of short-horns; and if I know it, I will not sell any to anyone who has not the same object in view. On this principle I began breeding, and I am convinced I have a better breed of short-horns in my possession at present than has been for the last fifty years in the best days of the Messrs. Colling.

"The bull you ask me about sending you, Duke of Northumberland, is everything I can wish in a bull, and Short Tail has taken after 2nd Hubback, his dam (Duchess 32nd), two crosses by 2nd Hubback, and Short Tail's sister, the best animal in my possession, I expect is in calf to the Duke of Northumberland. The six from which your two were taken were good, but the breed of the years 1835-6 were far superior to those six, though very good. Brokenleg (Duchess 34th), I offered you at 100 guineas. If you were to send twenty times that sum for her, and her produce, I would not take it now. You will remember I told you after buying the two heifers, that if either of them died on the passage, or did not breed when you got them home, I would give you the two nearest in blood to them. Now a sister in blood to your Rose of Sharon (calved since you were here), has produced a heifer to her sire Belvedere; and for the two I would not take 1000 guineas. These would have been yours now had yours not bred. I will not sell either cow or calf, but I have no objection to sell the bulls I breed from them, or from my Duchess tribe, which are far better animals than the Red Rose tribe; but I will not part with the females of these tribes at present."

Mr. Bates kept up a friendly correspondence with

many of the leading short-horn breeders in America, and many American gentlemen. Very many American agriculturists, who visited England, were guests at Kirklevington. Mr. Vail purchased a heifer called Hilpa from myself, and some animals from Mr. Bates, and my brother Robert Bell. I cannot do better than transcribe Mr. Vail's letter, in which their progress is detailed:—

"Troy, Sept. 27th, 1847.

"Dear Sir,—Your two favours of 15th July and the 9th August I received in due course, the former speaking of Annabella, and account of short-horns owned by Mr. Stephenson, and the latter giving a detailed account of the fairs (shows), which had then recently occurred in England, by which I am happy to learn the successful issue of the exhibition of your short-horns, and those of Mr. Bell, at Scarborough. I very much regret that the papers which you sent out to me, giving an account of this show, as well as the Royal Agricultural Society's show, have not yet come to hand. I am in hopes I may yet receive them. I cordially congratulate you upon your successful exhibition of your stock. I see it still maintains its usual ascendancy when exhibited for competition. These two letters I have delayed answering till after our New York State fair had been held, which took place, as I had previously informed you it would, on the 14th, 15th, and 16th days of September inst., at Seratoga Springs, about 30 miles from this place. The exhibition was about as usual, though not nearly so good as it would have been had it been held in this City, as it was off from the Hudson River, and at a distance from canal navigation, which prevented there being as many articles at the show, as there would have been, had it been on the line of our great thoroughfares. The exhibition of stock was, however, creditable to our State, and the attendance was numerous, especially of gentlemen of distinction, from most of the States in the Union, as well as many of the first men in the Canadas.

"And now you will allow me to say that while the blood of your stock maintains its deservedly high reputation at home, it is winning not less fame in America. I exhibited Hilpa, and the first premium in the first-class of Durhams was awarded to her, and the second premium was awarded to Mr. Prentice's cow, which I suppose is the best he has; he had two on the ground and in the same class. I sent my bull Meteor, out of Duchess and Wellington, up to the show for exhibition, only at the request of some friends, he having taken the State Society's first premium for best Durham bull in first class in 1844, and also first prize in first class of bulls of any breed at the same show. The bull Marion, owned by Messrs. Bell and Morris of New York, and which was bred by the late Earl Spencer of England, was on the ground and justly took the first premium in first class Durham bulls. The judges in their report on Durham bulls, made the following remarks, and which is extracted from their report:—'The Committee would mention here, as coming in this class, the justly celebrated bull Meteor, belonging to Mr. Geo. Vail, of Troy, which was on the ground for exhibition only; having taken the first prize at a former fair, was excluded from competing at the present. *We think he stands unrivalled.*' In the second class, or two years' old bulls, the first premium was awarded to Young Meteor, owned by Mr. Wakeman, and which bull I sold to him when about three months old. This bull has now taken three first premiums of the state society. He took the first prize of this society as best bull calf, and last year first prize as best yearling. He was got by Meteor and out of one of my Durham cows, but not from either of those I had from you. My two years' old bull also took the second premium in this class. Mr. Wakeman had also his two years' old heifer, got by Meteor, and her bull calf on the ground. This heifer also took the first prize as a two years' old heifer. He also purchased this heifer from me at the same time he purchased the bull calf, and this heifer has also taken the first premium at three successive State Shows as a heifer calf, a yearling, and now as a two years' old heifer, and her bull calf, about five months old, took the first premium as a bull calf, and one I exhibited took the second prize. I had four heifer calves on the ground, and one took the first and another the second prize, which were all the prizes offered in that class. I had no yearling bull on the ground nor was there any of my breeding exhibited, and Mr. Prentice took the first prize in this class. I had a yearling heifer on the ground; she was awarded the third prize. Geo. Ohlen took the first, and D. D. Campbell the second. The result is then, that in every class except one where I had an animal, or there was one of my breeding, they were awarded the first prizes. There was, however, a three years' old bull got by Wellington on the ground, which I sold when a calf with his dam, but he was poor in condition and not fit to show. He was not awarded any prize, and should not have been sent to the show in the condition he was in.

"I have thus been particular in detailing the result of this great show, as I doubt not you feel an interest in it, and that it will be highly gratifying to you. I suppose at this show there were thirty or forty thousand persons present, and many of the first men in the country, among them two ex-Presidents of the United States. As soon as our agricultural papers publish a full description of the show, I will send you a paper or two giving particulars. Hilpa has dropped a white bull calf, now about six or eight months old, got by Meteor; and Lady Barrington has also dropped a red and white bull calf, about four weeks ago, also by Meteor. One of these calves I have sold to go to Canada, at the close of navigation, at 300 dollars. The other I think I shall keep for my own herd. I have just got Wellington home, and think I shall put him to Lady Barrington and Hilpa, and hope to get heifer calves. Lady Barrington's calf last year by Meteor is a heifer, and a promising one. This is the only heifer calf I yet have had from the cows I have received from you, and hope I shall be more successful the next year in heifer calves.

Mr. A. B. Allen, of New York, whom you know, is continually urging me to get from you a young Duchess bull, and I would much like one, but at present I dare not venture the expense. Meteor, in some respects, is a finer animal than Wellington: he is better in the hind-quarters and across the hips. Wellington has not a broad hip, and is rather thin across the twist, and some of his calves partake of this defect. His fore-end cannot be beat; he is a superior handler, as also is Meteor. They both excel in this valuable quality. I weighed Meteor three days ago: his weight is 2,200 lbs., and Wellington, when in order, will weigh about 1,800 or 1,900. Meteor makes a splendid show, and, I doubt not, would take a high rank even in your country. Our county show took place in the city last week, and the best we have had. I was equally successful in winning premiums here as at the State Show. Hilpa took the first prize. A yearling bull which I showed, got by Meteor, and not from cows from you, took the first prize for the best bull on the ground; and my two years' old heifer and my yearling heifer each took the first prize in their respective classes. I only put one animal in each class, and took the first prize in all but one; and I believe this exception had the blood of my herd, and that of yours in part. Annabella, I hope, may make a good milker; she will never make a show cow; she stands entirely too high on her legs; her offspring may be better, as Wellington and Meteor are both right in this particular. You will see I have written you in a hurried manner. I could say much more to you, but must delay till I again have the pleasure of addressing you. With my most cordial wishes for your continued health, I remain, very respectfully, your friend, &c., GEO. VAIL."

"Thomas Bates, Esq."

To the above letter, Mr. Bates made the following memorandum, which was found among his papers after his decease:—

"The above is from the President of the Great American Agricultural Society, answering to our Royal English Agricultural Society in importance in that country. He got from me in the spring of 1840 his bull Wellington, out of my Oxford premium cow in 1839, and by my bull Short Tail, when he was only a few months old; and his Meteor bull was by his Wellington and out of a heifer he got of me by Duke of Northumberland. All the cattle he got of me obtained the highest premiums at their shows from their first going to America; and on Wellington bull being landed at New York, a person saw him, and understanding he had an elder brother, came over to England and told the gentleman who had bought the brother of me, that he had come over on purpose to buy him, and gave him all he asked for him—250 guineas—and returned with him to America, exhibited him ten times, was always successful, although their best bulls in America were brought out in competition against him. Hilpa was a cow bred by my tenant, Mr. Thos. Bell. "THOMAS BATES.

"Kirklevington, Oct. 31st, 1847."

Meteor's portrait was taken and engraved for inserting in Vol. III. of the Transactions of the New York State Agricultural Society, in 1843, to which was added the following note:—"Meteor was by Duke of Wellington, dam Duchess, both bred by Thomas Bates, of Kirklevington, Yorkshire, England, from whom Mr. Vail purchased them."

In June, 1848, Mr. Vail wrote to Mr. Bates:—

"I have recently purchased the entire herd of short-horns belonging to Mr. Prentice. It may, in the present state of demand for short-horns here, be considered by some as rather a bold operation. The herd consists of twelve head, being the four reserved cows which he considered the best of his herd, when three years ago he made his large public sale. These four and their offspring compose the purchase. Among this number there is a few animals which I conceive will breed well to my bulls Duke of Wellington and Meteor, which I propose to retain, and sell off the remainder. I have told the Messrs. Allan, and Colonel Sherwood, who have had the use of my bulls to their herds, that there was no other way to revive prices, so as to procure renumerating prices, for breeding than by introducing the blood of your herd, and through this medium to convince the public of its superiority, and I believe they concur with me in that opinion, though, I fear, they have not entire confidence of success, as I see they are making preparations to make sale of the most of their short-horns, at our New York State show, in September, at Buffalo. I do not myself like the idea at all of giving this matter up, as I conceive the short-horns are decidedly preferable, as a combination, for the dairy and the shambles, to any other breed of neat cattle. And at this crisis, unless some one or more do not persevere and keep up pure bred animals, the loss to the country will be incalculable. Now, what I want to aid me in my endeavours to sustain this stock, is a young red or roan Duchess bull from you, but I cannot, at present, think of increasing additional expense in procuring one. I have to-day received a letter from John Wittenhall, Esq., who, together with Hon. Adam Ferguson, of Nelson Gore, Upper Canada, purchased Lady Barrington's last bull calf, which he calls Halton, after the name of the County he represents in the House of Assembly, in Canada, saying he wishes me to have his name, pedigree, breeder's, and his present owner's names, recorded in the forthcoming volume of the British Herd Book, which will be of service to him and Mr. Ferguson. He was by Meteor, out of Lady Barrington 3rd, and is a red roan, and a fine animal, one which will do you much credit in Canada, as descended from your stock. The young bull has recently reached his destination, and the owners are much pleased with him, and intend to keep him as a close bull for their own herds.

"P.S.—I regret that the transactions of the last year's operations of the New York State Agricultural Society are not yet out for last year. I will by first opportunity send it. It contains a good portrait of Hilpa and her bull calf."

It is often mentioned as a remarkable fact that the Kirklevington herd in America produces a much larger number of bulls than heifer calves. Such calves being nearly all kept for breeding; the blood is thus spread over a large extent even of that vast country. The improved value of such a great quantity of stock soon covers any sum, however large, that is paid for the pure blood.

PART XI.

Durham Agricultural Society—Weighing and Measuring Cattle—Tyneside Society—Working Classes: attempts of the Duke of Northumberland to Improve—Crowley's Crew—Manufacturers supporting their own Poor and Schools—Success of Mr. Bates—Envy—Way to manage him—Short-horns in Northumberland—The Messrs. Jobson—Training for Shows—Booth and Crofton—Ruinous consequences—Mr. Bates again exhibits—Yorkshire Society, 1838.

The proceedings of the Durham Agricultural Society had especial reference to the killing value of cattle and sheep, and probably their experiments on the relation of live and dead weights might be utilized at the present day. I am informed that on the Continent the usual plan in the large markets, and at distilleries and sugar manufactories, where large quantities of cattle are fatted, is to sell cattle by live weight, the buyer of course exercising his judgment as to the quantity of dead meat he may expect, in the same manner that a miller in buying wheat by weight calculates what quantity of flour he may obtain. Mr. Mason made many experiments by weighing the food, and his stock at stated periods, and also on the different sorts of food.

Mr. Bates never had a weighing machine for his cattle, but he had tables for ascertaining the weights by measurements. These tables he obtained from his father, and they were compiled from experiments made by him, the Messrs. Culley, and other agriculturists.* These tables were often copied, and much used in the district. The measurements, in many respects, shewed the progress of the improvement of the stock, perhaps more pointedly than by weighing. The experiments at Halton, in feeding were often superintended by young gentlemen, who resided there for the purpose, and Mr. Bates in his letter to Professor Jamieson, in 1812, mentions two gentlemen who would be at Halton as companions to Mr. Mackenzie.

The Tyneside Agricultural Society, so often mentioned by Mr. Bates, had the support of all the principal noblemen and landowners of the district. The labourers and poor were not overlooked by the society. I may here

---

* Rules and Measures are now made from these Tables. Ewart's I have found most useful, and easily managed. See also C. Hillyard's letter on this subject, Royal Agricultural Society, vol. 3, page 337.

relate what was formerly attempted for the benefit of the working classes. The Duke of Northumberland (the Earl Percy of 1783) was patron or president of this society. This nobleman was most energetic in the improvement of his estate, and the tenants and labourers of all classes. He made most patriotic endeavours to reform the whole of the agricultural labouring population. The cottages on his estates were let with the farms and occupied by the hinds, *i.e.* married farm servants hired by the year, and each found a bondager, *i.e.* a woman to work for the farmers when required. His Grace took all the cottages away from the farmers and let them direct to the occupiers, adding half an acre of good land near each cottage for garden, and fixed the rent of both at 25 shillings a year; no rates were then paid by cottagers. Each occupier who could purchase a cow had sufficient grass land to graze and grow hay for it at a very moderate rent. The liberality of His Grace, however, proved generally a curse instead of a blessing to all the parties. The farmers soon found that they could not rely on the independent cottager for regular labour, and hired single men and women to reside and board in the farm-houses, and the cottagers and their families sought employment at manufactories and new works, often at great distances from their residences, but they clung to their cottage and potato ground with true Irish love, and generally with the same improvident result. The welfare of their working men was the first consideration of the founders of most of our large manufactories.

Sir John Crowley (the Sir John Anvil of Addison's Spectator), in the last century established at Winlaton and Swallwell the first large ironworks in the North of

England. His workmen were collected from all parts of the country, but principally from Yorkshire, Warwickshire, and Staffordshire. At first they were a most disorderly race, and in the county were called Crowley's crew. This worthy and benevolent gentleman made rules for regulating wages and disputes among themselves.

"He also made an excellent regulation for supporting their own poor, by each workman contributing eightpence a pound out of his earnings, which raises a fund that enables them to allow a man and his wife seven shillings a week when past work. A widow is allowed three shillings a week, and one shilling a week for each child. To entitle them to this privilege they must live all their time with the factory; if they leave it they lose the benefit of this association.

"He also established schools at Winlaton, Winlaton Mills, and Swalwell, for the sole benefit of the workmen's children.

"The proprietors built a chapel at Winlaton, capable of holding upwards of 300 persons, and pay the officiating clergyman.

"They also keep and pay a surgeon for attending the workmen whenever they require it."

Very many men brought up and educated under these rules became managers and owners of large establishments, and their descendents are now among the most eminent and wealthy in the Northern Counties. Similar rules in modern times might have saved much agitation about education and poor laws, and kept the land free from the charge thereof.

I have referred to the proceedings of the Tyneside Agricultural Society, and I find the name of Mr. Bates as gaining the prizes for bulls, cows, tups, ewes, the best swine, the best road mare, and also for turnips,—Sir Wm. Loraine, Messrs. Gibson, Jobling, Wailes, Harbottle, and numerous farmers also exhibiting and obtaining prizes.

The swine of Mr. Bates were a very superior breed, which I have continued, and now possess.

The success of Mr. Bates and the general state of his stock at Halton, no doubt caused great envy and jealousy, as mentioned above. I here quote the written account of Mr. Bates himself:—

"I exhibited cattle at the Tyneside Agricultural Society's meetings, from its first institution in 1804 to 1812, but never showed my best cattle, and was successful at every show (and these exhibitions were held sometimes thrice a year) till the last show in 1812, and I then showed better animals than I had ever showed before. One of the judges told me afterwards, that those who influenced the proceedings, decided in the room, before they went out to examine the stock, that I should never have another premium in that Society, however excellent the stock I exhibited; and this gentleman, and others who knew the determination come to, advised me never to exhibit again, and for 26 years afterwards I never did till the York meeting in 1838, although I continued my subscription to the Tyneside Society till it was dissolved.

"In 1819, seven years after I had ceased to show, a breeder of short-horns, removed to the northern part of Northumberland from Tyneside, dined with me at the same table at Berwick-upon-Tweed, and after dinner he asked me, before a very large company, how my stock was going on, and I said I had not exhibited since 1812, and knowing the resolution that was come to, 'that I was never to have a premium, however excellent the stock I exhibited,' he then said 'You ought not to reflect upon me seven years afterwards, for I avow myself to have been the proposer of that resolution;' and he set to work to defend himself and those who acted with him, 'justifying their conduct' by saying, 'that they were perfectly right in so doing, as they had none of my blood, and having, for so many years, given me premiums, it was time to put a stop thereto, and act for their own interest alone, and help the sale of their own stock;' this ended two years afterwards in the breaking up of the Agricultural Society in Tyneside; and the short-horned cattle of that district, from having been the best shows I ever knew, far exceeding any in the present day, as a whole, became the worst of any district I know of; for in 1837, at the Hexham Show, there was not even the vestige of a good short-horn from Tyneside in Northumberland, and the premiums were nearly all carried away by strangers from other districts; and with the decline of good short-horns, the agricultural produce of the district fell off to less than one-half to what I had known it on many farms, and probably never again will become equal to what it once was, while in other parts of the kingdom the agricultural improvements have greatly advanced. This ought to be a warning to all other agricultural societies to prevent the conductors thereof being governed by selfish motives, to advance their own interest, instead of the public advantage; for I have held, and ever will hold, that the prosperity of the landed interest, I mean landlord, tenant, and labour conjointly, tends to the prosperity of every other class in the state."

Mr. Bates wished the society to try experiments on feeding and test the stock, and offered to be at the sole expense.

I have taken great pains to ascertain in Tyneside, whether Mr. Bates was not under some misapprehension as to his treatment by this society, and a gentleman of high position, who was then a member of the society, informed me that Mr. Bates was quite accurate, and that he was present when the resolutions were come to, that Mr. Bates should have no more prizes. I said was this not wrong and unfair to Mr. Bates. He replied "No, not all; it was the only way we could manage him. He got all the prizes." This society was dissolved in 1821. The number of members had much diminished, and a great change for the worse had come over both the cattle and cultivation of the district. After discussing what were the real objects of such societies, they were satisfied that they had seen the cultivation of the land and the breeding of stock carried to a point with which they could not hope to improve by any exertions of the society, and therefore dissolved it. The Northumberland Agricultural Society was subsequently established, and also a local society in Tyneside by Mr. John Grey. Mr. Bates, however, always took a great interest in the agriculture of his native county, and at the meeting at Alnwick in 1840, his bull 2nd Duke of Northumberland, then let to Messrs. Jobson, took the first prize. Mr. Bates was present at this meeting, and at the dinner his health was proposed by the chairman, and drunk with great enthusiasm; for which he returned thanks saying that, "although he resided then in Yorkshire, his heart was still in Northumberland."

Messrs. Jobson were old friends of Mr. Bates, and they possessed the descendants of Jolly's Bull, as before mentioned. "Jobson's old sort" were long spoken of, but they had used bulls with Punch and Ben blood and their herd had much deteriorated. It however re-established its character after the use of the 2nd Duke of Northumberland. Mr. Bates agreed to take two heifers for his use for three years, or £50 a year, at the option of Messrs. Jobson. He tried to rouse the breeders in Northumberland to energy and stop the often-lamented decay of the breeds in that county, and he made most liberal offers of this bull for use in Northumberland, which, however, were not responded to. Fortunately this is now remedied by the spirited enterprise of Sir Walter Trevelyan and Sir John Swinburne, and others, whose herds contain many animals of Kirklevington blood.

The Trevelyan family have long been patrons of agriculture. Arthur Young records in his Northern tour the pleasure he derived from his visit to Wallington. The irrigation of Somersetshire was introduced in many places in Northumberland. The Messrs. Culley at Wark, and at Halton Castle and Ayden Castle, it was in use many years, but all the works are now abandoned, and are a puzzle for the antiquary. The Devon cattle could not compete with the short-horns and kyloes.

The late Sir John Swinburne was also an active member of the original Tyneside Society.

2nd Duke of Wetherby, and 13th Grand Duke (21850), are now in use at Wallington and Capheaton. The Princess of Yetholme, from the herd of Sir Walter, obtained, in 1863, three prizes as the best short-horn cow and, in 1869, the first prizes at Birmingham and Smithfield.

Lady Florence, in the Capheaton herd, obtained six prizes and one cup in Yorkshire, but will not be shown again.

Mr. Gow (of Cambo) has several first-rate animals, also of the Kirklevington blood.

Before referring to the exhibitions of the Kirklevington herd for prizes, I make a few observations to show the exertions of Mr. Bates in endeavouring to make shows really effective for the purposes for which they were intended. His labours met with no thanks or reward in his lifetime, and his motives were never appreciated. We have seen what his opinions were on this subject as early as 1807, and I think we must now admit that he showed not only patriotism, but wisdom and foresight in his writings at that period. I cannot more clearly express the opinions of Mr. Bates on feeding for exhibitions, than in the language of the historian of the Booth herds. Speaking of the management of cattle prior to 1838, he says :—

"At that period there were happily no shows to demand the sacrifice of the best cattle in the kingdom, as the few that were held could be reached by the majority of cattle attending them only by such long journeys on foot as would be impracticable by animals in such a state of obesity as is now a *sine que non*, with the judicial triumvirate. High feeding at that time meant no more than good pasture for cows early dried of their milk; and the term 'training' was never heard except in relation to horses. The first breeder who introduced the system, which has since run into such ruinous excess, of house-feeding cows and heifers in summer on artificial food, was Mr. Crofton; and in that year he, of course, took all before him in the show-yards."

Mr. Dixon says* "the Booth family began with Teeswater and twin brother to Ben (660), and lengthening the hind-quarters and filling up the fore-flank, and breeding with a view to that fine deep flesh and constitution, *which bears any amount of forcing*, have been their especial aim. It was the late Richard Booth's opinion that no bull had done their herd so much good as Albion (14), of the alloy blood, and Mr. Whittaker and Mr. Wetherell were quite with him on the point. It may be said that short-horns generally have grown smaller in frame, and that there is perhaps not that rich coat and uniformity of character

---

* Saddle and Sirloin, by the Druid. London, 1870, (page 150). The Druid was Mr. Henry H. Dixon, the author of the Prize Essay on Rise and Progress of Short-horns, published in the Journal of the Royal Agricultural Society of England, 1865. Vol. I., 2nd series, page 317. I fear I have quoted Mr. Dixon before without this explanation.

which invested some of the earlier herds, but still those who can make the comparison from memory are fain to allow that in their flesh points and general weights, the breed shows no decay. What the Brothers Colling were in earlier days the Brothers Booth have been in the latter. If the elder could boast of Necklace with the wondrous crops, and Bracelet, in whom none could find a fault, save a trifling deficiency in the fore-rib, it was left to the younger to keep up the type with the beautiful Charity, &c. Richard Booth and Crofton might be said to have *initiated the modern plan of keeping beasts far more in the house and preparing them especially with a view to shows.* No blood has been more widely spread than that of Warlaby and Killarby throughout the United Kingdon, or commanded *a finer bull-hiring trade, &c.*"

Mr. Carr, in reference to the state of fat in which breeding animals are exhibited by short-horn breeders, and after enumerating the number of prizes obtained by the Booth Herds at each exhibition, says:—

"It is idle, however, to condemn those who adopt this ruinous system, so long as the judges award their prizes and the public their commendations to animals in this unnatural state. *The breeder must follow the fashion*, or be left behind in the race, as systematically as the ancient husbandman selected from his herd this to propagate the breed, and that to bleed a victim at the sacred shrine, *must the modern short-horn breeder, who would maintain his position before the world*, yearly single out the choicest of his herd for immolation on the altar of the Royal Agricultural Moloch. Nevertheless, hard as it might be with such cattle to forego the triumphs of conquest, *good sense* would seem to say—if such be the cost at which renown is to be purchased—

"' Then rather let my herd, as leisure leads,
Wanton inglorious o'er the grassy meads.'

I hope to satisfy the short-horn breeders that Mr. Bates had the courage and also the good sense not to follow this ruinous fashion, and that to him alone the world is now indebted for the preservation of this old valuable race of cattle in a pure and uncontaminated state, and I hope the sacrifices and exertions of Mr. Bates to accomplish this will be duly appreciated.

Mr. Carr again on this fatal training says, (page 93):—

"Mr. Booth deprecated the system, yet was obliged in great measure to confirm to it. Though his cattle were absolutely unrivalled in their aptitude for quality and ample development on pasture, and he repeatedly sent them in blooming health and burly condition from the pasture to the show field, and occasionally with success, he too frequently on such occasions came off second best; hence it was found necessary to subject, for an adequate period, such of them as were designed for exhibition to a system whose disastrous effects, even the vigorous condition of his cattle were unable finally to counteract; hence the failure of female representatives of the Blossom and Charity tribes, &c."

It is difficult to gather amidst the poetry the eloquence of Mr. Carr's history in what the Booth blood really consisted. They had no tribes which can be traced or numbered like the Kirklevington herd, and I have the impression that he has proved that all the old valuable blood of the Booth family, what they had even in the days of Colling, has entirely disappeared in the female line, and that the stock which now boast that name are all descended from the alloy bulls Albion and Pilot, and the market cows.

Of the herds of the Messrs. Booth, Mr. Carr has so clearly and concisely written that I will say no more. They were honourable, upright men of business. They carried on the "Bull Hiring" Trade. "But though Mr. Booth deeply regretted the necessity of shewing his cattle, he still felt compelled to do so, even to the last, and he shewed Prince of Battersea, at Newcastle, only a few months before his death, &c. But Booth blood was fated still to maintain that proud *premium* which it has always held when fairly tested. Animals of kindred blood, in the language of the turf, 'took up the running.'" Mr. Carr also enumerates the prizes obtained by the alloy cattle of Colonel Townley, Mr. Ambler, Lady Pigott, &c., "and how they conquered in a hundred fields." Messrs. Booth had to sell and let their stock, and to do so must follow the fashion of the day, whatever that might be. This no doubt they did with great success and honour. Charles Colling had obtained the highest prices for his cross bred or alloy cattle. Why should the same not be continued, and the alloy blood continue to preserve its superiority in appearance and price? Alloy bulls and market cows can produce animals, one-half of which can be trained for exhibition, and the other half sold as stock, and so represent the *elite* of the short-horn race. But then, to keep this up, bulls of *sterling blood are necessary*. Where are they to come from? The bad handling of the O'Callaghan cattle, and of the Johanna tribe, surely did not represent the merits of the ancient class of short-horns.

It was an unfortunate day for the comfort and peace of mind of Mr. Bates that he ever consented to exhibit his herd for public competition. He was as it were an amateur, having to meet and contend with professionals. He says he was urged to show the world what pure bred improved short-horns were by a gentleman who accidentally saw his herd. I believe myself that his great anxiety was to show young breeders, and especially the Americans, what improved short-horns really were. Mr. Bates would not follow the fashion. He would not bow down to or worship the images of short-horns set up. Not even those of the Royal Agricultural Society of England. But the neglect or rejection of his cattle was a matter of public or national importance in his eyes. It was not a personal one.

Mr. Bates did exhibit at York in 1838 on the formation of the Yorkshire Agricultural Society. It was with great reluctance he gave up his resolution never to exhibit again. He sent seven animals to contend for eight prizes.

1.—The prizes were for the best bull of any age £25; second ditto, £10. 15 competitors. First Premium to Hecatomb, the property of Earl Spencer; second ditto, Mr. Wiley.

2.—For the best two-years-old bull, £15; second ditto, £5. 9 competitors. First Premium to Thomas Bates; second ditto, Mr. Linton.

3.—For the best yearling bull, £10; second ditto, £5. 9 competitors. First Premium to Mr. Wiley; second ditto, Mr. Childers, M.P.

4.—For the best cow of any age, in calf or in milk, £10; second ditto, £5. 9 competitors, 12 animals. First Premium to Mr. John Colling; second ditto, Mr. Edwards.

5.—For the best cow, three-years-old, in calf or in milk, £10; second ditto, £5. 4 competitors, 6 animals. First and second Premiums to Thomas Bates.

6.—For the best two-years-old heifer in calf, £10; second ditto, £5. 7 competitors, 10 animals. First Premium to Mr. Bates; second ditto, Mr. Edwards.

7.—For the best yearling heifer, £10; second ditto, £5. 7 competitors, 8 animals. First Premium to Mr. Edwards; second ditto, Mr. Bates.

Mr. JOHN GREY, Dilston,
Mr. BENN, Lowther Castle, } Judges.
Mr. SMITH, West Rayson,

So that Mr. Bates obtained five prizes and lost three.

On the health of the successful candidates being proposed at the dinner by P. B. Thompson, Esq., Mr. Bates acknowledged the toast in a very animated speech, in the course of which he alluded to the condition of cattle breeding when he first knew it fifty-four years before. The interest of agriculture he considered of permanent national importance. It was a science to which chemistry and all other branches of natural philosophy ought to give place. He alluded to the proposal for introducing a section of agriculture in the British Association, and regretted that there was no professorship of agriculture in either of the universities. His address elicited the warmest applause.

Lord Dundas congratulated the meeting on the success likely to attend the society. He lamented with Mr. Bates that there was no professorship of agriculture at either Oxford or Cambridge.

PART XII.

Mr. Bates and Judges at Shows—American Opinions and Sympathy—John Grey on Judges—Arthur Young's Farming—Mr Bates's Rejected Cattle—Oxford Show and Victory—Road and Sea Journey there—Daniel Webster—Political Farming and Banking—Idea on Exhibiting Families of Cattle—Cleveland Lad—Cambridge Meeting—Opinion Vindicated at Liverpool—Hull Meeting and Success—The Victory at York over the Booth Best—Necklace and Bracelet—Their Pedigrees and Victories by Carr—Ceased to Exhibit—Fat Breeding Stock and its Results—Calves like Cats.

Mr. Bates was never a judge at any show, but often, when requested, he nominated or recommended judges, and he often objected that the judges were selected from breeders of short-horns, or who were interested in them.

I may here state that no record was kept of the exhibitions of the Kirklevington herd, and I have now great difficulty in ascertaining what animals were actually exhibited. Although I daily saw the herd, and managed them on all occasions, I cannot recall all their movements. I have had to obtain my information from the published reports of the various shows, and the letters of Mr. Bates, so that some inadvertent errors may have crept into my account of the animals exhibited. Mr. Bates always kept a regular Herd Book, in which he entered the service of the cows, and the birth of the calves, and sometimes afterwards an entry of the names. No account was kept of the sales or hirings.

The Yorkshire Society was formed in 1838, and its first meeting was at York. Mr. Bates hoped that the evils and dangers of exhibitors, which he had pointed out in his letter of 1807, would have been guarded against in 1838, and that, under the presidency of Earl Spencer, true bred improved short-horns would have been duly appreciated and judged of. The result of the York meeting, and the subsequent proceedings of the Royal English Agricultural Society, kept him in a turmoil to the end of his life. He exhibited until he had vindicated the character of his herd, and also his own judgment, and then retired and left the arena to the alloy and fashion. He remonstrated and made his last offer in the cause, in 1848, as will be seen afterwards.

The only thing which consoled and supported Mr. Bates was the opinion and sympathy of his American friends, and the appreciation of pure bred improved short-horns in America. One American gentleman, of great experience and judgment, told him, after having spent a long time at Kirklevington, and examined every animal, "I left America very much prejudiced against short-horns. I had heard much of your herd, but I could not believe that it could be such as it was represented. I can now say it far exceeds what I had heard. I could not have believed that there was such a herd, unless I had seen it. Yours is the only stock that can improve the American breeds." In like manner another American gentleman, considered one of the best judges in that country, when he returned home, incurred great opprobrium when he stated, "that the Kirklevington blood was the only one which could benefit the Americans."

Mr. Bates was not satisfied with the results at York. He thought he ought to have had the prize in Class 1 with the Duke of Northumberland, who got the prize in Class 2, and that he ought to have had the prize in Class 4, and that he ought to have been 1st in Class 7, and also had both prizes. Perhaps the judges thought that he should not be allowed to get "all the prizes." Mr. John Grey evidently knew that Mr. Bates was not satisfied, and in responding to the toast of the judges, enlarged on the difficulty they had in pleasing everyone. Mr. Grey, on a subsequent occasion did but give satisfaction, and he mentioned this difficulty in a letter after a Highland Society meeting, in which he says,*

"The dissatisfaction expressed with the decision of the judges at Glasgow will make it more difficult to get men to act in that

* Memoir of John Grey (page 325).

capacity in future, and the apparent disinclination of the great landowners to come into contact with the great body of tenantry, by whom it is observed, will have an unfavourable effect on the future of the Society, &c.

"P.S.—I made some remarks at Glasgow, in the end of my speech, on the natural partiality of parties for their own stock, and some excuse for the fallibility of human judgment. It may also be observed, that lookers-on, or those who examine cattle in pens, have not the same means of comparing the quality or handling that the judges have, when they are brought side by side. And that is a point to which much attention should, in my opinion, be given, for a form, however correct, without good quality of flesh, is not a desirable animal."

I crave the indulgence of my readers in calling their attention to this show. The Duke of Northumberland's history I will state hereafter. One at least of the cows rejected founded the tribes of the Cambridge Roses. One judge, and he might be in the minority of three, admired the rejected ones. Mr. Grey was no doubt a man of honour and acted to the best of his judgment, and Mr. Bates never charged him with anything except a want of knowledge of short-horns. I hope I have written nothing which could be in any way considered as any imputation on his character, or the purity of his mind and life. Mr. Grey came into Tyneside after I had left. In my recent visit to Northumberland, I made enquiries respecting Mr. Grey's short-horn knowledge. I knew that Mr. Bates in 1842, in writing about the portrait of the Duke of Northumberland and his dam, said that Hecatomb's portrait in the Herd Book was a faithful likeness, as all Mr. Davies' were, but he was not satisfied with the portrait of the Duke of Northumberland, and regretted that it should have been inserted in the Herd Book. He wrote this of the two portraits. "I do not expect any artist can do them justice. They must be seen, and the more they are examined the more their excellence will appear to a true connoisseur; but there are few good judges. Hundreds of men may be found to make a Prime Minister for *one* fit to judge of the real merits of short-horns."

I fortunately met perhaps the only gentleman now living (to whom I related the great dispute) who knew the old class of short-horn breeders. He was a political friend and great admirer of the character of Mr. Grey. He said however, that, "Mr. Grey never was a judge of short-horns. He knew very little about them, and he succeeded better with sheep." Mr. Grey's sheep, however, when brought by him to Tyneside, had none of the properties or grandeur of the old race of Leicesters of the Messrs. Donkin, Bates, Jobling, and others. Mr. Donkin's sheep are engraved and described in Bewick's History of Quadrupeds. This gentleman added, "Arthur Young was always well known as the worst practical farmer in his county." I heard also much regret expressed that the biographer of Mr. Grey should have represented that all the improvements in the condition and agriculture of Northumberland also, were due to the success of the Whig and Radical political programmes, abolition of slavery, Catholic emancipation, the advent of the Whigs to office, and the Reform Bill, Free Trade, Tenants' Rights, Ballot, &c.

Mr. Grey became and was the only representative of a numerous class in Tyneside who had flourished and disappeared without leaving any representative behind them, and he no doubt did his utmost to resuscitate the former agricultural spirit, but I fear he found very few to assist him of the class I have mentioned in my earlier pages. His biographer and the public labour under the mistake that what he wrote and spoke about owed its existence to him. He was a tenant of Earl Grey, and for his political services to the Whigs, obtained an office of £1200 a year; and for a political appointment, it was a fortunate one for the tenants and the estate, as it relieved them from the necessity of publicly competing for their farms.

I also heard much sympathy for Mr. Grey, and regret expressed, that confiding in political and religious creeds in trade transactions, he had unfortunately been involved in a bank, where the Quakers managed with more success than in Overend and Gurney. They converted an insolvent bank into a joint stock company, and had sufficient time to pay large dividends out of capital and to sell their shares and be all free from liability before the crash came.

Mr. Bates himself in relating what occurred at York, stated that, when the judges had concluded their decisions, one of them who stands as high as most men in public estimation, and with whom I had been intimate for a long period, came to me and said, " Will you oblige by accompanying me to look at three animals which you have exhibited this day, for I consider them the best three animals I ever saw in my life." I did accompany him, and he went first to the Duke of Northumberland, and pointed out all his good qualities. He then took me to Red Rose 13th, which gained the highest premium at Cambridge in 1840. He next took me to the Duke of Northumberland's own sister, Duchess 43rd. Now, this same judge, in conjunction with his brother judges, had, *in all these cases*, placed two animals before each of them, and in every instance these lacked all the good qualities that this judge pointed out that mine possessed. I have stated that the cow which gained the premium at Cambridge was one of them, and when shown at Cambridge, she appeared far inferior to what she did at York, having only some weeks before had a calf taken from her by a severe operation, from which catastrophe she never afterwards perfectly recovered, and lost many stones in weight before the show. I need not state the merits of the Duke of Northumberland as the judges placed him first in the two-years-old class. The Duke's own sister, that was placed nowhere at York in 1838, gained the highest premium at the Oxford Meeting in 1839, as the best year old heifer, and at the Yorkshire Agricultural Society's Meeting at Hull, in 1841, was placed first as the best three-years-old cow, whilst the one placed second at York, in 1838, was Duchess 42nd as a year old, and was placed second to Duchess 43rd, in 1841, as a three-years-old cow; this being the second time I had obtained first and second premiums for three-years-old cows,—in 1838 and 1841. As I had commenced exhibiting cattle again, and been so improperly treated by having my best three cattle placed nowhere, I determined that they should see *the rejected ones* again, and in every instance where they were a second time exhibited they compelled the judges to do them justice; and the same judge who coincided with his colleagues at York, in 1838, came to congratulate me that autumn on my great success at Oxford. I said "I sent my rejected cattle at York to Oxford. When the Duke obtained the highest premium as the best bull at York, in 1842, one of the judges who decided against him in 1838, as being the best bull of any age, was compelled to assent to his being placed first."

At the above show at York, Mr. Bates had his seven animals placed together for examination as above mentioned.

The chief breeders pressed upon and examined them, and this no doubt suggested to Mr. Bates the idea which he afterwards carried out, that short-horns ought, at some part of the exhibition, to be shown in families, and which he impressed upon his American friends. I think it is not quite accurate that all the rejected at York were sent to Oxford. The Cambridge Rose was intended to go, but was not in a fit state to travel, and the Matchem Cow, which had got a prize at York as second to a cow of his own, went. The Duke and Matchem Cow, and the two Duchess heifers, one red and the other roan. (Which form one picture). Mr. Bates travelled to and from London in the steam-ship from Middlesbrough with his animals, and himself saw to their treatment. The Duke of Northumberland in landing in London slipped and lay across the gangway, but Mr. Bates patted his head and called him "poor boy," and the animal remained quite motionless until he was rescued, and, happily, he received no injury. They travelled by road on foot to Oxford, and Mr. Bates returned from London with them, but on landing said he would never again go by sea. I believe it was his only sea voyage.

The Royal English Agricultural Meeting held at Oxford, 1839, Earl Spencer, President.

SHORT-HORN CATTLE, Class 1st.   Judges: Mr. T. Charge, Mr. W. Smith, and Mr. J. Hall.

To the owner of the best bull, 30 sovs., 7 competitors. *Mr. T. Bates*, for his bull aged 3 years 9 months, prize. To the owner of the best cow in milk, 20 sovs., 4 competitors, *Mr. T. Bates*, for his cow aged 4 years 8 months, prize. To the owner of the best-in-calf heifer, not exceeding 3 years old, 15 sovs., 3 competitors. *Mr. Thomas Bates*, for his heifer aged 1 year and 11 months, prize. To the owner of the best yearling heifer, 10 sovs. Competitors—Mr. R. W. Baker, Mr. G. Davey, Mr. G. Carrington, jun., Mr. J. H. Sangston, Sir George Philips, Bart., Mr. Wm. Warner, Sir Charles Morgan, Bart., the Marquis of Exeter. *Mr. Thomas Bates*, for his heifer aged 1 year 10 months, prize. To the owner of the best bull calf, 10 sovs., 4 competitors. The Marquis of Exeter's bull calf aged 7 months, prize.

The *Farmer's Magazine* thus remarks:—

"Although the exhibition of stock, was not as numerous as had been expected, yet the quality of some was very superior. The animals shewn by Mr. Bates were universally admired as excellent specimens of the short-horn breed; and Mr. Paul's Devons excited general admiration."

[At the Yorkshire Agricultural Meeting, held at Northallerton, 1840, the prizes were:—

For the best cow in calf, or in milk, of any age, £20. First, *Mr. T. Bates*, for his Oxford, calved in 1834; second, £10, Mr. J. Booth, for his Yorkshire Jenny, calved in 1836.]

At the grand dinner which took place in the quadrangle of Queen's College, Mr. Daniel Webster, the celebrated American orator and statesman, in his speech, said:—

"Gentlemen,—In the country to which I belong, societies, on a small scale like the present, exist in many parts, and they have been found to be very highly beneficial and advantageous. They give rewards for specimens of fine animals, and the improvement of implements of husbandry, which may tend to facilitate the art of agriculture and which were not before known; and they turn their attention to everything which tends to improve the state of the farmer, and I may add, among other means of advancing his condition, that they have imported largely to America from the best breeds of animals in England, and from the gentleman who has been so fortunate as to take so many prizes to-day; from his stock on the banks of the Ohio and its tributary streams, I have seen fine animals raised which have been supplied from his farms in Yorkshire and Northumberland."

Mr. Bates sent only three animals to the Cambridge show in 1840, viz.: Cleveland Lad, the Red Rose, rejected at York, and a bull calf afterwards called Duke of Cambridge. He would not risk exhibiting his best females, and Cleveland Lad was not trained.

The Royal English Agricultural Meeting, held at Cambridge, 1840, Duke of Richmond, President.

SHORT-HORNS, CLASS 1st. Judges: Mr. John Wright, Mr. Eaton Clarke, and Mr. Wm. Smith.

To the owner of the best bull, calved previously to the 1st January, 1838, 30 sovs. Competitors—the Rev. D. Gwilt, Mr. Adeane, Mr. J. Bedsly, Mr. Barnard, M.P., Rev. E. Lindsell, Mr. T. Bates. Mr. Paul, for his 4 years 3 months bull, prize. To the owner of the best bull, calved since the 1st January, 1838, and not more than one-year-old, 15 sovs. 12 competitors. Mr. R. M. Jaques, for his 1 year and 10 months bull, prize. To the owner of the best cow, in milk, 15 sovs. Competitors—The Earl of Hardwicke, Rt. Hon. Charles Arbuthnot, Mr. Henry Beauford, the Marquis of Exeter, Mr. Jonas Webb. *Mr. T. Bates*, for his 6 years old cow, bred by himself, prize. To the owner of the best in-calf heifer, not exceeding three-years-old, 15 sovs. 5 competitors. Right Hon. C. Arbuthnot, for his 2 year and 3 months old heifer, bred by himself, prize. To the owner of the best yearling heifer, 10 sovs. 12 competitors. Mr. R. Jaques, for his 1 year and 10 months heifer, bred by the Earl of Carlisle, prize. To the owner of the best bull calf, 10 sovs. Competitors—Rt. Hon. C. Arbuthnot, Mr. J. S. Tharp, Mr. John Warsop, Mr. Barnett, Mr. R. B. Harvey, Mr. Carrington, jun., Marquis of Exeter. *Mr. Thomas Bates*, for his 8 months old bull calf, bred by himself, prize.

Mr. Bates was not satisfied with the rejection of Cleveland Lad. This appears in his subsequent letters. He also objected very much to the decision of the judges on the animals in classes where he did not exhibit, and to the fat and trained condition of the whole of the animals.

Cleveland Lad was the only animal shown by Mr. Bates at the Royal Agricultural Society's meeting at Liverpool, in 1841, and at the council dinner, Philip Pusey, Esq., M.P., presided, and Earl Spencer occupied the vice-chair. After the awards of the judges were read, the chairman rose and said :—

"Lord Spencer and Gentlemen,—I have now the honour of proposing to you the health of a gentleman who has distinguished himself as a competitor for stock this day. Though not one of the best judges of stock myself, I know enough about it to be aware how much has been done by the Royal Agricultural Society of England for the general benefit of the country. (Cheers.) I know, as I dare say all of you do, that there are certain proofs of the excellence of stock, and I know that very great improvements have been made, through the efforts of this society and those of the Smithfield Club, in the breed of our stock. I am unable to state what are all the recent improvements which has been made." He then reviewed the old accounts of cattle in a humorous speech, and quoted the author who says, "That in those days England was well known for surmounting other countries in the breed of cattle, as may be proved with ease, for where are oxen commonly more large of bone?" (Laughter.) He then proceeds, "In most places our graziers are now grown to be so cunning, that if they do but see an ox or bullock, and come to the feeling of him, they will give a guess at his weight, and how many score or stone of flesh and tallow he beareth, how the butcher may live by the sale, and what he may have for the skin and tallow— (laughter)—which is a point of skill not commonly practised heretofore. Our first improvement in cattle was in the long-horned breed, and I believe they are now little seen. There was one long-horned beast at the last Smithfield Show which was much looked at. There are few, I presume, to whom we are more deeply indebted for improving the roast beef of old England than the subject of this toast. Gentlemen, I beg to propose to you the health of the successful competitor in Class 1, Mr. Bates of Kirklevington, Yorkshire." (Applause.)

Mr. Bates rose and said, "he begged to return his most sincere thanks, on behalf of himself and the successful candidates, for the honour they had done them. He congratulated the assembly on the glorious prospects which were held out as the results of that Association to the whole world; for he felt convinced that by the improvement of agriculture they affected the interests of the entire people of the earth. (Hear, and cheers.) They had arrived at that glorious era when the welfare of every man was dear to his fellow-creatures; and he hoped the Society would do its utmost to encourage that feeling. (Hear, hear.) It was by that alone England had risen so high in the estimation of the world. It was by that alone she could rise still higher; and he also congratulated them on the efforts now making by their Society to secure that end. He only wished those efforts would be increased a thousandfold."

The Yorkshire Agricultural Society at Hull, 1841.

Judges for Cattle—Mr. Hunt, Mr. Geo. Davidson, Mr. J. Wright.

CLASS 1.—For the best bull of any age. 21 entries. *Thomas Bates*, Kirklevington, for his bull 4 years 7 months old, the winner at Liverpool. Prize, £30. For the second best bull—R. M. Jaques, for his bull Clementi (well known.) Prize, £10. James Watson's white bull of Wauldby, Hull, was greatly admired, as were many others. The judges commended the whole of the stock shown in this class as very good.

CLASS 2.—For the best two-years-old bull. 10 entries. R. M. Jaques, for his bull Clementi, the winner in the 1st class. Prize, £20. For the second best two-years-old bull—William Brandham, £10. The judges spoke in commendation of the other animals in this class.

CLASS 3.—For the best yearling bull. 5 entries. *Thomas Bates*, Kirklevington, for his bull Duke of Cambridge. Prize £20. For the second best yearling bull—J. M. Hopper, of Newham, Stockton-on-Tees, for his bull Newham, £10. The judges commended the whole of them, especially the bull shown by Mr. Booth.

CLASS 4.—For the best bull calf. 12 entries. Jonas Whittaker, Burley, Otley, for his bull calf Lord Adolphus Frederick, £10. Second best bull calf—Earl Spencer, Wiseton for Zenith, £5. The bulls in this class were generally commended by the judges.

CLASS 5.—For the best cow of any age, in calf or milk. 9 entries. John Booth, of Killerby, Catterick, for his cow Bracelet, 4 years old. £20. For the second best, John Collins, for his cow Madame Grisi, in calf, 10 years old, £10. The cows in this class were generally commended.

CLASS 6.—For the best three-years-old cow in calf, or in milk. *Thomas Bates*, of Kirklevington, for his cow by Belvedere, own sister to Duke of Northumberland, £15. For the second best cow by Belvedere. *Mr. Bates*, £5. Mr. Tor's cow, of Rigby, Caistor, was commended.

I believe that Mr. Bates did not exhibit in Class 5.

In 1841 the career of Mr. Bates may be said to have ceased. He had vindicated the character of his herd by exhibiting the rejected animals, and in every case had been successful. He was not satisfied with the proceedings of the Royal Society, as I shall mention hereafter.

Mr. Bates had always been on most friendly terms with Mr. John Booth, who frequently visited at Kirklevington, and was always a guest at the Yarm Fair, in October. Mr. Booth took especial delight in bantering Mr. Bates about short-horns, and his declining to exhibit at the Royal, and joked that he dare not show a cow, and if he would, he (Mr. Booth) had a rod in pickle for them. This referred to Mr. Booth's cow, Necklace. That my readers may clearly understand what the fear was, and what Mr. Bates had to meet, I extract Mr. Carr's account:—

"Toy was a very neat, thick framed cow, with a magnificent udder. Her milking capabilities were the boast of Killerby dairymaids, and were transmitted to her famous twins—Necklace and Bracelet. The latter were bred from very close affinities; thus Vestal, the dam of Toy, was by Pilot; Argus, the sire of Toy, was by Young Albion out of Anna by Pilot; Toy's two daughters, Necklace and Bracelet, were by Priam, a grandson of Pilot on his dam's side, and whose sire, Isaac, was by Young Albion, from Isabella by Pilot. Toy had twice previously been put to Young Matchem, but the offspring, Teetotum and Plaything, were not at all equal to the twins. Teetotum, however, was the parent of Lady Thorn by Lord Stanley, whose portrait in the Herd Book—evidently a portrait—unmistakeably proclaims her a good cow, and in truth she was second only to Birthday in merit. She was sold to Mr. Banks, Stanhope, for 150 guineas, a great price in that day, and in 1845 won the first prize at the Royal Exhibition at Shrewsbury in the cow class; her two daughters, Ladybird and Revesby Thorn, subsequently earning fame in the show fields. From Toy was also descended the cow Gertrude, purchased together with her lovely daughter Lady Hopetown, at Mr. Bolden's sale, by Mr. Torr, who had previously bought, at the Killerby sale, Sylphide by Morning Star, also sprung from Toy. Some others of her descendants are in the hands of Mr. Pawlett, and Mr. Torr's noble bull Breastplate has well kept up the honour of the family. But I must here return to the twin born progeny of Toy, the all-conquering Necklace and Bracelet. Necklace and Bracelet shared the pasture and strawyard with the ordinary stock of the farm until nearly two years old. As calves they never had more milk than their dam (who suckled them both) supplied; and throughout the whole of their victorious career, they derived their chief support from the pasture, with a daily *bonne bouche* of corn and cake. Yet Bracelet won seventeen prizes at the various meetings of the Royal Agricultural Society of England, the Highland Society of Scotland, the Yorkshire Society, and other local shows; and at the Yorkshire Show in 1841, where she won the first prize for extra stock, the sweepstakes for the best lot of cattle not less than four in number, was awarded to Bracelet, Necklace, Mantaline, and Ladythorn. Necklace won sixteen prizes and one gold and three silver medals at the various meetings above mentioned, as well as at the Smithfield Club, where she finished her career as a prize-taker in 1846, by winning the first prize of her class and the gold medal (for which there were thirty-seven competitors), as the best animal exhibited in any of the cow or heifer classes. At the Smithfield Show, in the following year, the same prizes were awarded to Mr. Wiley, and in 1849 to Mr. Cartwright, for animals bred from the Killerby stock. In five years four first prizes for the best short-horn cows at the Royal Agricultural Society's meetings were awarded to animals bred by Mr. Booth, of Killerby, in 1841, at Liverpool, to Bracelet, in 1842, at Bristol, to Necklace, in 1844, at Southampton, to Birthday, and in 1845, at Shrewsbury, to Ladythorn."

"To this day it is a moot question amongst those who remember the world-renowned twins, to which of them could be justly awarded the palm of beauty. Necklace is said to have had neater forequarters, and to have been rather better filled up behind the shoulders. Bracelet had fuller, longer, and more level hind-quarters. Bracelet was the dam of the famous bull Buckingham, by Musselman, and of Morning Star, by Raspberry, which was sold in 1844 to Louis Phillippe. She also produced Birthday by Lord Stanley, whose career as a prize cow at the Royal, the Yorkshire, and the County of Durham Shows, was eminently successful. Bracelet was also the dam of Hamlet, by Leonard; and whilst in calf with him, her stifle joint was dislocated, and being incapacitated for further breeding she was slaughtered."

It required no ordinary courage to enter the list with such a heroine as Necklace. She could boast of the blood of Pilot on Albion, alloy on alloy. But passed over the blood of the Irish O'Callaghan, and the thick skinned hornless poor Scotch relation. To meet this cow, supposed to be the best the Booth herd ever produced, Mr. Bates had his broken legged cow, Duchess 34th, that had never been exhibited. She broke her leg when about a year old. Mr. Bates would not employ a veterinary, and I myself, with the assistance of the journeyman miller at Kirklevington, set her leg. She was consequently kept in the folds for nearly a year, and made no growth during that period. Brokenleg was in milk, and had pastured in the usual way with the other dairy cows, since the 12th of May she had no *bonne bouche*,* whatever that may mean, of either corn or cake. She had no training whatever, and walked by road, about 40 miles, to York, with her son the Duke of Northumberland. Mr. Bates sent the Duke to this show, as he had only got the two-year-old prize in 1838. Of course he got the 1st prize, and thus completed the list of the recoveries. Necklace came with colours flying, covered with prizes and medals won in many a show field, and in full training from her Bristol victory. I was not present at the show at York, and Mr. Bates did not return home for some days after. I therefore had the account of the show from old "Tommy Myers," the cowman. When he arrived home, by the road, I asked him how they had used him, he replied, "vary weel, but I thout at yan time they were gannin to use me vary badly, at first we walked the fifteen coos altogether twice round the ring, and then they ordered me to stand at one side, where I was for half-an-hour or so, and then they ordered two others out, and keepit disputing ower them. I thought they were settling which was to be 1st and 2nd, but they at last sent me the white rose, and wasn't I pleased." The dispute had been about the 2nd prize, between Mr. Mason Hopper's cow, and Mr. Booth's, but it was at length awarded to Necklace.

* I asked a gentleman to translate this. A farmer, who was present, said he could not translate it, but that it meant "belly measure," as much as they could eat.

Mr. Bates received a letter from a valued friend, dated January, 1847, in which he said:—

"I was in London at the Christmas show at Smithfield, and a competitor at the show with an ox. He was, however, not fat enough, although he was one of the first, if not the first beast sold, proving that the butchers did not consider him a bad one. The judges could not have given him a premium for fat. The feeding was certainly excessive. Some of the animals were said to have been fed upon sugar; and, although I think feeding is all fair enough at Smithfield, we ought all to feed alike. You are aware that Necklace, after having received premiums at Newcastle and Wakefield, got the gold medal at Smithfield. I do not think she was the best beef, although a good cow: but it proves what the state of the animals is at *Breeding Shows*. They are in fact fed until they will not breed. I understand Bracelet's last calf was not much larger than a cat; she dislocated her stifle joint and was slaughtered. It is a pity to feed animals to such an excess as to destroy them. They cannot go on breeding."

PART XIII.

Royal Society and Breeding Stock—Rules Neglected—Pandering to the Masses—Wonderful Thanks and Credit to Mr. Booth—Mr. Bates' Exertions—Judges at Shows—Mr. Bates protests—Red Tape—Letter of Appeal to the Members of the Society—Views adopted by Americans—Booth Beef *versus* Milk—Mr. M'Combie, M.P.: his opinion—Best Cattle not now Exhibited—Neglect of Duchesses 34th and 51st—Foreigners Misled by Shows and Training—Its effect—Holland Cattle—Short-horns in Germany—Mr. Carr's results of Alloy-blood—Poor Relations, &c.

The Royal English Agricultural Society in their "Instructions to Judges" say:—

"As the object of the society in giving the prizes for neat cattle, sheep, and pigs, is to promote improvement in breeding stock, the judges in making their award are requested not to take into their consideration the present value to the butcher of the animals exhibited, but to decide according to their relative merits for the purposes of breeding."

In noticing the last public acts of Mr. Bates and his exertions in the cause of short-horn breeding, which he made at York in 1848, I am much relieved by the admissions and confessions made by the Royal Society, and my readers may judge for themselves how the society acted up to the above instructions, from their Cambridge meeting.

In the report of the Plymouth Meeting of the Royal Agricultural Society, published in the *Journal* in 1865, by the senior steward of the yard, J. Dent Dent, Esq., he says:—" The yearling bull class was decidedly better than the older bulls." The judges say of the yearlings:—

" We had some very superior animals in this class. The bull which we placed first was very far above any of the others. The one placed second, Mr. Booth's Commander-in-Chief, from his age and condition, is not calculated to please the *multitude who are not in the habit of seeing animals in store condition.*" Mr. Dent continues: " I may here say that whatever be the difference of opinion as to the merits of the animals, Mr. Booth and his late uncle deserve credit for showing Commander-in-Chief at this show, and Prince of Battersea at Newcastle, in useful working condition and not overlaid and disguised with fat; and some courage on the part of judges is required to recognise the merits of an animal which is only in useful, not in show condition."

If Mr. Bates had been alive he would probably have expressed his thankfulness that there is a Republic in which short-horns, in a store condition, are allowed to appear in public. My agricultural friends must be prepared for the next sop to the multitude, which will be that the cattle at shows are to be immediately consumed, and therefore only fat ones permitted to be shown. With this dread of the masses before them, Mr. Bates certainly showed fool-hardy courage when he undertook his last enterprise to establish breeding cattle as an institution, if not in the state at any rate, in the Royal Agricultural Society of England's show yard. His estimate of the greater difficulty in obtaining judges than Prime Ministers, is probably not exaggerated when in addition to the skill, courage to face the multitude is also required.

That my readers will calmly judge whether credit ought not to be awarded to Mr. Bates, and also a monument erected to the courage that he so long and perservingly displayed in 1848 on this subject, which appears to have so bewildered the Royal Society. In general, people stood aghast and trembled when they saw Mr. Bates questioning the decrees and practices of the Royal Society. The multitude no doubt considered the Royal Society infallible. How could any one doubt the wisdom, or criticise the acts of a Society, with such an array of royal, noble, and honourable names—to say nothing of the crowds or multitudes who annually attended their gathering. The chief towns in the Kingdom vieing with each other for the honour of a visit, and raising unheard of sums to tempt their visitors by sumptuous preparations and entertainments.

Mr. Bates could not discover any member of the council who was a judge of short-horns, and he was not the only person who questioned the judgments of the Royal Society. I have recently discussed the subject with several gentlemen in Northumberland, and they say that in their opinion, and in that of their fathers, who had seen the old stock of short-horns, at the Newcastle Meeting, in

1846, the best animals did not obtain the prizes, and that the prize short-horns were not pure bred, and shewed their want of breeding.

The Americans always coincided with, and adopted the views of Mr. Bates, and it was by their knowledge and judgment that he was supported, and aroused to action. I believe I cannot more clearly express what Mr. Bates did and said, in support of his recommendation that the stock should be in a breeding condition, and that pedigree should be regarded, than by inserting his letters, omitting such parts only as are a repetition of what has been before written, or which are too personal, although, probably only too true and pertinent. I have not copies of the letters (no doubt they contained unpalatable truths) written by Mr. Bates to the society. But I have copies of three, the first addressed to Lord Yarborough the president, and when that was not noticed, a letter very similar in its statements to the members of the Royal Society, and the third to an American gentleman.

"To the Members of the Royal Agricultural Society of England."

"On becoming a member of the Society in 1838, I sent the Society a copy of a letter I had addressed to the Board of Agriculture, &c., in 1807, on the importance of improving the live stock of the country, hoping that my views might have been acted upon, and at the first country meeting of the Society at Oxford in 1839, I exhibited four short-horns, and was again an exhibitor at Cambridge in 1840, when I found that great offence was taken at my being again an exhibitor, and had I not said at the council dinner, the day before the exhibition, that I would exhibit the bull then rejected at Liverpool, I should not have again exhibited. Cleveland Lad, now the property of Lord Feversham, where he has been six years, and is above twelve years old, and yet is likely long to be useful as a bull.

At Liverpool he received the highest premium, although the bulls shown at Liverpool were superior to those shown at Cambridge.

"I did hope when Lord Portman had given notice in December, 1847, that he should make a motion respecting the choosing of judges of the Society, that men of real judgment would have been chosen and not men totally ignorant of what constituted a good short-horn, or else persons connected or partners with the exhibitors; and I did hope when the show-yard was to be opened after the judges had concluded their decisions, by persons paying one pound each for admission, that great good would have resulted therefrom, and I wrote several long letters to the Society pointing out the propriety of showing the various herds together, as it could then be done on the evening before the exhibition day, when they must stand as placed for the public to see them when classed. And I wrote three letters to the Society, offering to send four generations of short-horns for exhibition, provided the Society would agree to allow the judges to look at the four generations, before they made their decisions. To these letters I received evasive answers, which caused me to write three times for explicit answers, saying I would not send any cattle but on that condition, the answers were, 'they would refer my letters to the stewards of the yard.' The time was come then that I must give the prescribed notice or I could not show, and I gave the required notice, but entered my four best animals as extra stock, not for premiums, that the less offence might be taken was my motive for not exhibiting for premiums, and when the numbers to be placed on each animal before they entered the show-yard were sent me, I gave the generations I had sent distinguished by the numbers only, that the judges might not know whose they were, or how they were bred, certificates of which I sent both the Secretary of the Society and the Steward, that both might know every particular relating to those cattle, and begging that the same might be entered on the catalogues of the show by itself, at the end, so that the public might examine over the four generations where each stood, which the numbers showed. I also particularly requested that I might then be informed whether the judges would be desired to examine the four generations, and if not, that I would pay the penalty for cattle not sent, but to this I received no answer.

"Now all my cattle were kept in the ordinary way—the cows going at grass night and day, as well as the young stock, not kept on hay and cake, and forced forward in condition, as had been done by the prize cattle at York, and mine were without turnips the previous winter months, not having turnips for them.

"When I arrived at York I found they had selected one good judge, but he was to act for Devons, Herefords, &c., and not for short-horns.

"When notice was given that persons paying one pound each might enter the yard, as the judges had concluded, I entered, but no one was there; no catalogues, no decisions made known, nor any person who was connected with the society or stewards of the yard, &c.

"After a long search I found the men with bulls, and then learnt that the judges had never examined the four generations of short-horns I had sent. That there were only two judges for short-horn bulls, and that they had never handled one of the bulls; and that of the two judges only for female short-horns one was a partner with the greatest exhibitor.

"As a meeting of the Society was to be held the next morning, the only time a member of the Society can attend, or be allowed to speak, I attended, hoping to be able to prevent such conduct in future, when it had just occurred, and had met with universal disapprobation, not only from members of the Society, but from foreigners, who knew good English short-horns better than most English breeders. On meeting with one from America, he immediately said, 'The cattle to which the premiums for short-horns have been awarded are not short-horns.' And he knew my cattle by name, though he had never seen them before, and was just arrived for the first time in England a few days before. The president would not allow me to speak or make any remarks, and said I might write to the council. In order to do so I asked for copies of my letters, but the secretary said in reply, 'that the council regretted that they could not supply me with copies of my letters.'

"Deeply I regret it, but in all probability the best short-horns will go abroad and may never return to their native land till the present generation die out, when their successors may be glad to obtain *the finest variety of cattle that ever existed.*

"Had the Royal Agricultural Society of England allowed the judges to examine my four generations of cattle at York, and admitted of a discussion on the subject, I was ready prepared for the purpose, and should have laid before the society my long experience and its results, to benefit breeders of short-horn and other cattle. I would have thought it incredible had I not the fact from an observer, that any two men, as at York, would have decided on the merits of cattle without handling them, as was done with the short-horn bulls. Now it was by handling that Mr. Charles Colling knew the merits of Hubback bull, and thus preserved the ancient short-horns, to which I have invariably kept close, and thus alone have I been able to produce my herd while all the other breeders by neglect thereof have lost that on which merit depends.

"Now had the judges at York been accustomed to handle cattle, they would have *handled*, and such gross decisions would never have been made; and as was observed at Cambridge by a gentleman, who had seen my cattle here. 'Had I been (he said) blindfolded I could have known all your cattle by the feel of my fingers,' to which I replied, 'As the Stewards of the Yard have heard your just remarks I hope in future the judges may be blindfolded.' Had men judgment—and without handling they cannot have it—coarse ill-bred brutes, which are by forced keep made fat, would be rejected at once, and then quick grazers, *however lean, when exhibited*, would have a chance to gain prizes, but while size and fat are alone regarded, no right decisions will ever be made.

"The reduction of the money prizes given by agricultural societies to one-tenth of what is now given, would be amply sufficient now that railroads convey the stock gratuitously. Animals would then be more likely to be shown without the cost of feeding. It was an empty compliment paid to me by a steward of the yard when he said, 'Your stock were the only ones shown in a proper state of condition,' when invariably the premiums are given for size and fatness—*quality totally disregarded* by the judges at the public shows; although they say in the instructions to the judges the very contrary to their invariable practice.

"Having shown the two Cleveland Lad Bull's sister (Oxford 2nd) and her four youngest calves—two bulls and two females—and a grandson of hers at the Yorkshire Agricultural Meeting at Scarborough, in 1847, *and obtained prizes for all the six.* My 2nd Duke of Oxford being placed before Capt. Shaftoe* that had gained the highest premium at the Royal Agricultural Society at Northampton the previous month, and knowing well that a numerous and powerful party use all their influence to prevent, if possible, my cattle gaining premiums at York, and glory in their having accomplished their object, the public cannot expect that I will again be an exhibitor. And had even my last letter to the stewards been acted upon not one wrong decision would have been made, for when the person did act, whom I had recommended, not one decision was wrong made in 26 different lots, and when finished, he and the other two acting with him were desired to examine and decide on the short-horns exhibited as extra stock,—not shown for premiums—and they highly commended my two short-horn Duchess cows, and also commended the two heifer calves I had sent, and the officials of the Royal Agricultural Society have put the cattle shown only as extra

---

* This bull's history was a less chequered one. Mr. Unthank had become deeply smitten at Richmond with his short legs, rich quality, and gay looks, when he was the first-prize yearling of the Yorkshire Agricultural against Belleville, Cramer, and Belted Will; but there was no little difficulty in persuading Mr. Lax to part with him for £200. The late Mr. Benn, always enthusiastic in the short-horn cause, lent the Lowther van, and as Mr. Unthank sold The Captain after a couple of seasons for a £100 advance to Mr. Loft, of Lincolnshire, his second Richmond thoughts proved as good as his first. Mr. Parkinson, of Leyfields, gave the last bid of 325 guineas for him at the Trusthorpe sale, and won in the aged class with him at the Northampton Royal, the same year that his half-brother, Baron Ravensworth gained that honour among the yearling bulls. After coming second to Mr. Bates's First Duke of Oxford, at the Yorkshire show of that year, he changed hands a fourth time, for 140 guineas, to Mr. Smith, of West Razen, who kept him for five years, and then sold him to his brother, in whose hands he died. He had a great propensity to fatten, and got his cows very good and compact, but rather too small.—(*H. H. Dixon*, page 98.)

stock in their annual report amongst the rejected shown for premiums, so that those who read their report may consider such commendations as amongst those rejected and not thought worthy of a premium by their judges, when in fact, they were never exhibited for any premium."

It may be asked how it comes to pass that breeding stock cannot be exhibited for prizes in a breeding and natural condition? Is it true, as Mr. Bates asserted, that the judges generally appointed at the large shows do not know how to appreciate stock, and therefore give the prizes for beef. Mr. Bates would not sacrifice his herd to be in the fashion and obtain prizes. He did not wish to let or sell one-half of his herd by sacrificing the other half in beef, and therefore he did not exhibit. But when he did not exhibit, and had no interest in the animals shewn, yet he gave his opinion on the stock exhibited, and he says that very frequently, in nearly every case, at the Royal Agricultural Society, the judges selected the wrong animals as improved short-horns.

We have also the protest of Mr. Booth against the fat exhibition. Mr. Booth, however, as a man of business, had to follow the fashion. He had to please the public and attract customers. Mr. Booth was guilty of no deception. Short-horn breeders, in purchasing from him, knew that Albion and Pilot were not pure-bred improved short-horns; and if the alloy or half-breds, with their bad handling and want of quality, were fattest, and brought the highest prices at C. Colling's sale, in 1820, why should they not continue to do so, and breeders and the public award them their honours and praises? They gave little milk, but Mr. Thos. Booth pointed to their broad backs and exclaimed, "Look there, that is worth a few pints of milk." The same may be said of many other celebrated cross-bred cattle. Mr. Bates did not deprecate cross-breds. I find in a letter to Mr. Bailey, in 1810, written after Mr. Bates had read the sheets of the Durham report, when Mr. Bailey says that Charles Colling bred from the best kyloes, but gave it up, finding the pure short-horn superior, Mr. Bates requests Mr. Bailey to correct this, as Charles Colling had frequently told Mr. Bates that he could never procure the best sort of kyloes, and Mr. Bates had presented to Mrs. C. Colling one of his, which he obtained with such care as above appears, and that a heifer from this cow was then in calf, and C. Colling, after his sale, refused 100 guineas for her (she is mentioned in the Herd Book as bought of Mr. Bates) and that Mr. C. Colling said he never had a better grazer than this cross-bred heifer.

The opinion of Mr. Bates, respecting the Galloway red cattle, is confirmed by that of Mr. W. McCombie, M.P. for Aberdeenshire, and the celebrated breeder and feeder of the cattle of that county, in his interesting lecture on cattle and cattle dealers, which he printed at the request of the Scotch breeders, in which he places the Galloway cattle at the bottom of the list of the Scotch tribes, and among the slow feeders and grazers.

Mr. Bates always maintained that the agricultural exhibitors, and the show herds of those in the bull-hiring trade, gave foreigners a false impression of the improved short-horn cattle. The Americans soon saw the truth, and were not misled, although the pleasing of the multitude has in some recent cases tempted them to give large prices for show animals.

The Royal Agricultural Society may have great meetings, and spend incredible sums in exhibitions, and keep up a corporation of noble and honourable names, and secretaries, officers, and publishers, and the most correct red tape rules, but, it is by the private enterprise and exertions of Mr. Bates, and his successors, that the world is indebted for the preservation of the improved short-horns. We know that all the exertions of the Societies to please or pander to the taste of the multitude *will not* destroy the breed, and that at any rate as Mr. Bates foretold, Englishmen may still obtain improved short-horn cattle from America.

I mention an instance of the effects of the neglect of its own rules by the Royal Agricultural Society. Mr. Edward Bates, a nephew of Mr. Bates, on the introduction of the learned Chevalier Bunse, the Prussian Ambassador in England, settled on one of the Royal Domains in Prussia. On entering on it in 1848, he took the stock inventory, as usual, by valuation. There was a dairy of about sixty cows, some fine animals of the Swiss and Holland cattle, and several very mixed of the breeds of the country. He took bulls and cows the descendants of the old Dobinson and Brown herds, mentioned by Mr. Bates, which had been crossed with Colling's Meteor and the best bulls in Northumberland, and recently by Kirklevington bulls. Of course he used only the best short-horn bulls with all his cows, and had no doubts of the superiority of the best short-horns over all others, not only for feeding, but for dairy and draught. For several years his fat cattle sent to Newcastle were specially noticed in the market reports as being the best from the

Continent, and also in the market, and even in Smithfield they attracted a crowd around them. The oxen after being weighed were generally 100 Imperial stones, and obtained the prizes at the German shows. To test the value, and also to meet the assertions that the Kirklevington herd were tender, he bought calves from the best herds in the country, and treated them in precisely the same manner as his improved and crossed short-horns, and found the latter in every respect superior for work and for keeping in condition, and when fed and slaughtered the natives were about 25 per cent. of less value. He did not weigh the food, but had every reason to suppose that the natives consumed the most, and they were longest in being ready for sale. No fat cattle had been fed there; the farmers said they would not pay, and therefore they grew no green crops. Mr. Bates soon showed them that it would pay to feed proper cattle, and he grew large quantites of green crops (mangolds), and thus not only got the profits of the feeding stock, but obtained manure for his corn crops. The gentlemen who visited his farms had, many of them, been at our short-horn breeders' establishments, and cannot believe that the stock of Mr. E. Bates are the same race. It is difficult to convince them that improved short-horns need not always be in training, and that the blood of his herd is as good or better than that of the animals they see at shows and training establishments in England, and that training and showing cattle, and bull hiring, are themselves a business which it will not pay him to engage in. The prizes at the shows are too small, and the prices to be obtained are not sufficient to make it profitable to destroy the breeding capabilities of the stock, and to be always purchasing cows (as only men like Lord Althorp could afford) to keep up the herds. Mr. E. Bates has his cattle in good keeping and breeding condition, and when he wishes to sell them he can very rapidly make them prime fat and obtain the highest price per lb. for them. At the distilleries and sugar manufactories the well bred live cattle always command a much higher price than coarse ones. They pay due regard to quality and handle. Generally the attempts of landowners to introduce short-horns have been lamentable failures. They buy prize animals or animals from prize herds. They seldom breed and soon pine away if not kept on training food, and the broad backs are no recompense for the loss of pints of milk. Cows are milked thrice a day. Mr. E. Bates says he frequently buys English prize animals at butcher's price in Germany. The parties who have tried them became disgusted or tired of them, and will not continue to feed animals at the training rate with no prospect of repayment.

Mr. E. Bates has often purchased Holland in-calf heifers. The fashionable and almost universal colour in Holland is black and white, and the best are all of that colour, and it is fast superseding all others. In Friezeland and along the coast of Hanover, roan and red and white cattle exist, but they are inferior in quality to the black and white. The best Holland cattle from the neighbourhood of Rotterdam are very little inferior to the best improved short-horns. The cows have, many of them, all the qualities possessed by the old celebrated short-horns as described by Mr. Bates and others.

I cannot conclude this subject without mentioning

that Colonel Gunter, Sir John Swinburne, and other owners of first-class short-horns, decline to exhibit. They will not sacrifice their stock to please the multitudes, or compete with animals that, as Mr. Bates often said, had no value beyond the price to the butcher.

The Americans, with their usual sense and intelligence, have adopted the recommendation of Mr. Bates, and show their breeding stock in families. My readers will have observed that Duchess 34th and Duchess 51st were the two animals not thought worth a look, much less a touch, by the judges of the Royal Society at York, and the officials of the Society put them in the last sheet of the Journal. Mr. Etches and the Americans appreciated them, the latter crowded round them in large numbers during the exhibition. Probably, I am not wrong when I say that the Americans by their appreciation of the merits of the two cows, have proved by *The Times* test of auction, that they were worth more than all the female short-horns in the yard together.

Mr. Carr, in writing of the celebrated prize taking bull Julius Cæsar, says :—

"No matter how dissimilar and opposite in form and breed the cows to which he was put might be, the produce all bore the unmistakeable stamp of their sire. The offspring by him, of the shabbiest lane-side cow, had it is said, all the character of the pure-bred short-horn. It may be worth while to inquire how far the remarkable property which distinguishes this bull, may be traced to the preponderating influence of any particular progenitor or progenitors in his pedigree, an investigation of which, it may here be sufficient to say, will show him to be descended half-a-dozen times, and some of them very often more from twin brother to Ben. This circumstance lends weight to the opinion of many experienced breeders, that in general the capability of a bull to transmit to his offspring his own peculiar mould and properties, depends upon his having inherited them from a succession of ancestors endowed with similar characteristics. It is doubtless to the concentration of hereditary force thus derived, that the extraordinary transmissive power of such bulls as Comet, Favourite, and Julius Cæsar, is to be attributed. At the same time it is a curious circumstance, and one that should not be forgotten, as often modifying to some extent the principle above enunciated, that amongst animals similarly bred, there are some bulls, and some cows too, that possess an immeasurably greater transmissive influence than pertains to others."

Now does this modification of the principle so clearly enunciated not arise from the fact that Julius Cæsar was not a pure bred short-horn, and that twin brother to Ben was not descended, as stated in the Herd Book, from Hubback, and justifies the following passage in a writer oft describing the success of alloy cattle in America. "Noteworthy also was the success of Rosedale and Baron Booth, of Lancaster, at the different fairs in the Northern and Western States, but the Kentuckians and Western Statesmen considered the Booth blood full of alloy, and that the few animals imported had not made much work, so they would not use it."

If we calculate the number of times any bull, such as Favourite or Hubback, occurs among the ancestors of any animal, we may have the chance of the off-spring taking after such ancestors. Mr. Carr gives, probably, a correct mode of making such a calculation, but he nowhere mentions that the chance of the Scotch Galloway, or any other inferior ancestors, may have of cropping up among the descendants. Poor relations, are not blazoned in the bovine peerage or reckoned; with cattle, as with mankind, they are overlooked and forgotten. Mr. Bates said one cross might be corrected or got rid of by proper company, but he never knew two that could be so treated.

PART XIV.

The Prize Stock of Mr. Bates's Breeders—Spraggon's Herd—J. Wilson's Prize Bull—Lord Althorp's Brindled Black Nose—Chrisp's Herd—Wynyard Herd—St. Albans—Princess Tribe by Wood's Herd degenerated—Pedigree made clear—Messrs. Stephenson, Greenwell, and Harrison, clear accounts—Belvedere purchased—Dukes 224 and 226—Sheep destroyed, and the Law—Trial of New Tribes—The six Kirklevington Tribes, &c., &c.

I may here remark that all the stock exhibited by Mr. Bates afterwards continued to breed. The Oxford and Cambridge tribes are descended from his prize cattle. The Oxford heifers were unfortunate, one broke her thigh and was slaughtered, the other died of the foot and mouth disease. None of his stock were destroyed by training. Although Mr. Bates did not exhibit again, stock with his blood, continued in his lifetime to distinguish themselves in all shows. I may add here that Edward Bates has invariably found that when he used the pure bred bulls, he never had any black and white calves from the Holland cows, but with bulls he has purchased they frequently occur. Mr. Spraggon, of Nafferton in Tyneside, for many years used Kirklevington bulls, and had Sir Robert that did great good in the neighbourhood. At his sale about 1858, Mr. Jacob Wilson, of Woodhorn Manor, Northumberland, purchased his celebrated prize bull Duke of Tyne for £14. I was informed that the dam of this bull was a white cow with a black nose, and that on the day of the sale she was carefully locked up. Mr. Spraggon, jun., however informs me that this cow was purchased at Lord Althorp's sale, and that she was a roan cow, but she had a brindled head and a black nose, and was a very fine animal. We have seen how the prophecy of Mr. Bates that the Earl would not get a black-nosed calf was fulfilled. The price of this bull, however, showed most clearly the opinion of the public, that the cow was not a pure bred improved short-horn, yet her produce, like the alloy, obtained the royal honours. Mr. Chrisp, of Hawkhill, Northumberland, used the Kirklevington bull Refiner in his herd. He had bred a bull which took all the honours of his year, but at his sale the stock from Refiner sold for the highest prices.

Nothing has puzzled and perplexed the short-horn world more than the history of the Wynyard short-horns. St. Albans from that herd, so mysteriously introduced at Chilton had regenerated that herd, and his good qualities have never been called in question. The partizans of pure blood and of alloy both admit that in this case pure blood did show its nobility. Belvedere, from this herd, has caused more controversy. The worshippers of the Royal Society and the owners of their honoured animals were equally sore when their honours were tarnished by the fact that the Kirklevington herd contained animals which had only to be brought out, even without training, to surpass them in the shows, while the public in Europe and America expressed their appreciation of the pure blood. As in Tyneside, this fact interfered with the bull hiring and selling trades.

It is not surprising, however, that every effort was

made to disparage the blood of Belvedere, and the alloy faction were loud in victory when they could claim a relationship with Belvedere, and that the Kirklevington herd were, to use the shepherd's word, " tarred with the same stick."

A writer, in 1858, says of the Princess family:—

"The pedigrees given of Princess are not only various but conflicting. At the sale of Sir H. V. Tempest's herd in 1813, she is said to have been by Favourite, her dam by Favourite. The same pedigree is given in the first volume of the Herd Book, published in 1822. The third volume of the same book, published in 1836, gives her pedigree as being by Favourite, dam by Favourite, 2nd dam by Hubback, 3rd dam by Snowdon's bull,—by Waistell's bull,—by Masterman's bull, by the Studley bull. This is to be found in the pedigree of her grandson St. Albans (1412). Again, where the pedigree of Belvedere (1706) is given in the same volume, she is said to have been by Favourite, dam by Favourite, 2nd dam by Hubback, 3rd dam by Snowdon's bull,—by Masterman's bull,—by Harrison's bull,—bought by Mr. Hall, of Sedgefield, of Mr. Pickering. Princess was, as we have always heard, an extraordinarily fine cow, and the late Sir H. V. Tempest, of Wynyard, in the County of Durham, who, as an admirable judge of the points and excellencies of the whole animal kingdom, from a race-horse or short-horn down to a game cock, resolved, it seems, on being her owner at any price. It was never known what he gave Mr. R. Colling for her, secrecy on that head appearing to have been part of the bargain; but it has been said to have been as much as 700 guineas.

"Princess and her numerous descendants were sold in 1813, after Sir Henry's death. The late Mr. J. Wood,* of Kibblesworth, bought one of her daughters, Nell Gwynne, then in calf to Wynyard, and she afterwards produced St. Albans. We believe the pedigree of

* I understand that this herd which had been continued by Mr. Wood's son was recently sold, and that the use of the best alloy did it no good. It had degenerated and became smaller, another proof that the males ought to be of the best blood.

St. Albans, and consequently of Princess above alluded to, have been furnished by that gentleman, and we are sure from our thorough knowledge of him, with the most perfect good faith. We cannot but suspect, however, that an error has crept into it, from a confusion of the pedigrees of sires with those of dams, Waistell's bull having been by Masterman's bull, Masterman's bull by the Studley bull. At this sale Angelina, the daughter of Anne Boleyn, another daughter of Princess, was reserved; and at subsequent sale a daughter of Angelina was bought by Mr. J. Stephenson, of White House, Wolviston, a neighbouring tenant farmer. From her descended Belvedere, who became the property of Mr. T. Bates, by whom we believe the pedigree of Belvedere, and consequently of Princess was furnished.

"We have heard Mr. Bates say, speaking of the pedigree of Princess, that such it certainly was, because her dam was own sister to Mr. R. Colling's White bull. Yet according to the pedigree of that bull, given in the Herd Book, he would have been own brother to Princess. Who shall decide when such doctors disagree? We feel, however, justified in assuming that Princess belonged to the class of those descended of cows belonging to the Collings, which had a pedigree before they obtained them, whether her first recorded ancestors was bought by Mr. Pickering, of Mr. Hall of Sedgefield; or was by the Studley bull, of whom they obtained her, it is not we believe known. Nell Gwynne, Mr. J. Wood's purchase, became as we have said, the mother of the well known St. Albans; and Mr. Mason, of Chilton, having obtained of him St. Albans' services for some time, he became thus as we have seen, the sire of No. 13 at the Chilton sale. The ancestress of Lord Spencer's Old Roguery, and her numerous progeny of Z's. Nell Gwynne had but one heifer-calf having descendants, this was by Layton (366). Mr. Wood sold it to Mr. Troutbeck, of Blencoe, in Cumberland, who had been his pupil. By Mr. Troutbeck this heifer was christened 2nd Nell Gwynne; of Nell Gwynne the 2nd, many descendants in a right line of females have of late years appeared in the market, and not a few have attained show-yard honours. At one time however, a notion existed that Mr. Troutbeck had not been very select in the choice of some of his bulls.

"From Angelina, the grand-daughter of Princess, descended Belvedere, obtained as we have seen, by Mr. Bates, from Mr. J. Stephenson, and to this cross with his Duchesses Mr. Bates always attributed the greatest advantages. Those who knew his worth knew also his strong prejudices, one of which was against the alloy, and a strong suspicion exists that Lawnsleeves ought to have had a place in the pedigree of Belvedere; but a belief having existed that, according to the Herd Book pedigree, Lawnsleeves had the alloy in it, he was accidently omitted. Some years ago the writer, when on a visit at Elmore, the well-known residence of the late George Baker, Esq., the owner of Lawnsleeves, discovered, on reference to dates, that Lawnsleeves' dam could not have been the daughter of the alloy George, but probably by Mr. Mason's George, who was much nearer at hand, that gentleman, moreover, being well known to Mr. Baker. Had this been but sooner known the discussion need not have occurred."

The pedigree of the Princess tribe of Mr. R. Colling had been obtained in 1820, from Mr. Hall by Mr. Bates for insertion in the Herd Book. This of course could have no reference to the purchase of Belvedere in 1831, but Mr. Bates was thus familiar with the breeding of the Princess (late Bright Eyes) race of cattle. Sir H. Vane Tempest was one of the leading agriculturists and stock breeders in the county.

The Wynyard agricultural gatherings are mentioned by Mr. Bailey in his "Survey of Durham," and were attended like those of Holkham, Woburn, and Workington, by all the leading agriculturists, and Mr. Bates was often present. The gathering in 1807 was attended by Messrs. Colling; and Mr. Bates, after the dinner, had a public discussion with them about the sale of Duchess and her daughter, and the giving up of the latter by Mr. Bates on a promise respecting the bulling of Duchess by Duke (224).

Mr. Bates saw and was familiar with the Wynyard herd and the Princess blood, and in 1820 had obtained all the particulars from Mr. Hall for the Herd Book. This herd had been sold in 1812, two months after the death of Sir Henry, with his farm stock, with no notice to the short-horn breeders. A few animals were reserved by the Countess of Antrim, and sold with her farm stock in 1818, in the same quiet manner.

Mr. Stephenson had long kept a small herd of short-horns. About 1830, Mr. Atkinson Greenwell, a cousin of Mr. Stephenson, occupied the Grove Farm, at Kirklevington, as tenant of Lord Falkland, and as Mr. Greenwell had kept a dairy of cows, and been interested in cattle all his life, he took a great interest and was constantly inspecting Mr. Bates's herd. Mr. Greenwell frequently spoke to me of Mr. Stephenson's very fine bull, and told me that Mr. Bates ought to buy him, as he was a fine roan, and Mr. Bates's nearly all red. I told Mr. Greenwell to name it to Mr. Bates. He said he had done so, but Mr. Bates would not listen to it. In 1831, Mr. Bates was disappointed respecting a bull, bred from his stock, which he had relied on, and Mr. Greenwell again urged him to buy Mr. Stephenson's, and said his grandam was bred at Wynyard. Mr. Bates said all the Wynyard herd was gone. He had never heard of the sale, and that he would give anything to obtain that blood, but did not know where to find it, and he supposed they were all extinct. At last, the mention of the Wynyard herd changed Mr. Bates's tone on the subject of Mr. Stephenson's bull, and he said he would go and see the bull, if Mr. Greenwell would go with him, and they accordingly went the next day. Mr. Bates having inspected the bull, Mr. Stephenson gave the pedigree which Mr. Bates wrote down, and he agreed to give Mr. Stephenson all he asked, namely, £50, not guineas. His pedigree being the same as given in the Herd Book, the only error as to Angelina, being that Angelina 2nd was the first and Waterloo the second calf from her, and this pedigree was forwarded to the Herd Book in 1835. Mr. Stephenson himself brought the bull to Kirklevington the next day, 23rd June, 1831, and received the cash for him. He was a man well known and always considered most upright, and he gave Mr. Bates the pedigree in his own handwriting, and signed by himself. Mr. Bates had never heard of Mr. Stephenson except from Mr. Greenwell, and he would never have bought a bull descended from Lawnsleeves or any animal with the alloy blood in him. Mr. Stephenson knew nothing of this feeling, and it is very improbable that he falsified a pedigree to deceive Mr. Bates for the sake of £50 for a six years old bull, and this being before it was supposed that the cross with Belvedere should produce such wonderful stock. The doubt in this pedigree was first raised by the alloy faction discovering that Mr. Harrison, of Forcett, had, in 1830, hired the bull Waterloo and issued handbills representing him as out of Anna, by Lawnsleeves. The pedigree had been sent with Waterloo but not in Mr. Stephenson's handwriting, and very soon after Mr Stephenson corrected it. Most complete evidence was obtained by Mr. Bates that Angelina, in 1815, was sent to Mr. Baker's, of Elmore, to Lawnsleeves, and had a calf called Anna in 1816, and could only be two years old at the time of the sale in 1818. Whereas the cow bought by

Mr. Stephenson had bred several calves and was at least eight years old, and that there could be no doubt that the cow purchased by Mr. Stephenson in 1818 was Angelina, and not the daughter, and also that Angelina was bred as stated in the pedigree.

"A gentleman wrote to ask me why Mr. Bates struck the bull Lawnsleeves out of Belvedere's pedigree? I could have answered the question but did not choose to do so myself, but through another channel. I therefore enquired of Mr. Greenwell, who accompanied Mr. Bates when he went to Mr. Stephenson's to purchase Belvedere, and he thus replied:—'I was along with Mr. Bates when he purchased Belvedere and heard Mr. Stephenson tell him how he was bred. I was also at Mr. Bates' next day when Mr. Stephenson brought the bull, saw Mr. Bates pay for him and get a receipt for the money, and also received a written pedigree signed *John Stephenson*, but no bull of the name of Lawnsleeves was in that pedigree, nor was Lawnsleeves ever once named between them either on that day or on the preceding day. And I am quite sure of this, that if the bull Lawnsleeves had a right to be in that pedigree neither Mr. Bates nor Mr. Stephenson were the men to strike it out.'"

Mr. Harrison's written statement is as follows:—

"In the year 1830 I hired the bull Waterloo of Mr. John Stephenson, of Wolviston, County of Durham. When the bull came to my house at Streatlam Grange, where I resided, there came with him a memorandum of pedigree. That memorandum stated that he was by Young Wellington, dam Anna by Lawnsleeves, g. d. Angelina by Phenomenon, g. g. d. Princess by Favourite, &c., &c. This memorandum of the pedigree of Waterloo was not in Mr. Stephenson's handwriting; from that memorandum I had some handbills and cards printed stating the pedigree as above, and advertising Waterloo to stand at my house to serve cows.

"At a time subsequent to the printing of these handbills in a conversation with him, Mr. Stephenson informed me that there was an error in the printed pedigree as I had published it, and stated, that Waterloo was out of Angelina by Phenomenon, and *not* out of Anna by Lawnsleeves. This was before the sale of Belvedere."

There is no record of the son of the Duchess of 1804, then two years old. Duke (224) in the Herd Book is that animal. The only notice is " Roan, bred by Mr. C. Colling, by Favourite." He was let or sold to Geo. Gibson, Esq., of Stagshaw Close House, near Corbridge, and must not be confounded with Duke (226), own brother to Duchess the 1st, bought by Mr. Compton at the Ketton sale. In 1804, Mr. Bates had a favourite ewe of the Leicester breed, upon which he set very great value, and had paid Mr. Donkin £10 for being sent to his tup; she had only produced tup lambs. Mr. Gibson's pointers being allowed to range by themselves, ran this ewe into the mill pond at Halton, and she was drowned when big with ewe lambs. Mr. Gibson at first declined to make any compensation, but finally the matter was referred and Mr. Bates obtained the value of the ewe at butcher's price, without any reference to her breeding or value of the pure breed. Mr. Gibson then refused to allow Mr. Bates the use of the bull, and he was used by every one else as a hack bull, and this was a subject of complaint by Mr. Bates against Charles Colling.

Duke (226) was afterwards purchased by Mr. Donkin, and is the animal referred to in the correspondence with Lord Althorp and Mr. John Grey, who put many cows to him.

Mr. Bates frequently bought cows and heifers both for himself and his friends, but they were not mere ordinary market cows. He would purchase nothing that had not the breeding properties of improved short-horns, and especially quality. Many such were then met with which no doubt had good breeding, although the owners could tell little about them, except some traditions that such and such bulls had been used or that they came from some person's stock, or a neighbourhood where good bulls were kept. He tried the effect (see the Phillippa, Brown, and other families in the Herd Book) of breeding from such cows when put to his best bulls, but was very often dissatisfied with the results and did not keep the produce for breeding. The pedigrees of his Duchess and Rose tribes, are given anterior to the time of the Collings. The other four tribes Waterloo, Oxford, Wildeyes, and Foggathorpe, were all descended from cows that came up to his standard, and were from owners and herds that had been long established and had crosses of the blood of the tribes which he was so anxious to preserve in their purity. I will at the head of each tribe give the account of their origin.

I think it unnecessary to give any statement respecting the cattle Mr. Bates had at different times. I only give the particulars of the six tribes to which he confined his herd for nearly the last twenty years of his life. The Duchesses are all descended from Duchess 1st, whose pedigree appears in that of the Duke of Northumberland (calved 1808), and I now trace the descendants of that cow.

	Colour.	When Calved.	Sire.	Dam.
Duchess 2nd	red & white	1812	Ketton 1st (709)	Duchess 1st.
Do. 3rd	do.	1815	do.	do.
Do. 4th	do.	1816	Ketton 2nd (710)	do.
Do. 5th	do.	1817	do.	do.
Do. 6th	do.	1819	Ketton 3rd (349)	do. 4th.

	Colour.	When Calved.	Sire.	Dam.
Duchess 7th	red & white	1820	Marske (418)	Duchess 3rd.
Do. 8th	do.	1820	do.	do. 2nd.
Do. 9th	do.	1822	do.	do.
Do. 10th	do.	1822	Cleveland (146)	do. 4th.
Do. 11th	do.	1822	Young Marske (419)	do. 5th.
Do. 12th	red	1822	The Earl (646)	do. 4th.
Do. 13th	red & white	1823	do.	do. 9th.
Do. 14th	do.	1823	do.	do. 6th.
Do. 15th	do.	1824	do.	do. 8th.
Do. 16th	do.	1824	do.	do. 3rd.
Do. 17th	do.	1825	The 3rd Earl (1514)	do. 11th.
Do. 18th	do.	1825	2nd Hubback (1423)	do. 6th.
Do. 19th	yellow red	1825	do.	do. 12th.
Do. 20th	red & white	1825	The 2nd Earl (1511)	do. 8th.
Do. 21st	do.	1825	do.	do. 3rd.
Do. 22nd	do.	1826	2nd Hubback	do. 9th.
Do. 23rd	do.	1826	The 2nd Earl	do. 11th.
Do. 24th	do.	1826	2nd Hubback	do. 6th.
Do. 25th	do.	1826	do.	do. 8th.
Do. 26th	do.	1826	do.	do. 3rd.
Do. 27th	do.	1827	do.	do. 16th.
Do. 28th	do.	1828	do.	do. 6th.
Do. 29th	do.	1829	do.	do. 20th.
Do. 30th	do.	1830	do.	do.
Do. 31st	do.	1830	do.	do. 26th.
Do. 32nd	do.	1831	do.	do. 19th.
Do. 33rd	roan	1832	Belvedere (1706)	do. 19th.
Do. 34th	red & white	1832	do.	do. 29th.
Do. 35th	red	1833	Gambier (2046)	do. 19th.
Do. 36th	red & white	1834	Belvedere	do.
Do. 37th	do.	1834	do.	do. 30th.
Do. 38th	roan	1835	Norfolk (2377)	do. 33rd.
Do. 39th	do.	1835	Belvedere	do. 30th.
Do. 40th	do.	1835	do.	do. 19th.
Do. 41st	do.	1835	do.	do. 32nd

	Colour.	When Calved.	Sire.	Dam.
Duchess 42nd	roan	1837	Belvedere	Duchess 30th.
Do. 43rd	red	1837	do.	do. 34th.
Do. 44th	red & white	1838	Short Tail (2621)	do. 37th.
Do. 45th	do.	1838	do.	do. 30th.
Do. 46th	do.	1838	do.	do. 34th.
Do. 47th	red	1839	do.	do. 37th.
Do. 48th	red & white	1839	do.	do. 30th.
Do. 49th	do.	1839	do.	do. 30th.
Do. 50th	white	1839	Duke of Northumberland (1940)	do. 38th.
Do. 51st	roan	1840	Cleveland Lad (3407)	do. 41st.
Do. 52nd	red & white	1841	Holkar (4041)	do. 38th.
Do. 53rd	roan	1842	Duke of Northumberland	do. 41st.
Do. 54th	red	1844	2nd Cleveland Lad (3408)	do. 49th.
Do. 55th	do.	1844	4th Duke of Northumberland (3649)	do. 38th.
Do. 56th	red & white	1844	2nd do. (3646)	do. 51st.
Do. 57th	roan	1845	2nd Cleveland Lad	do. 50th.
Do. 58th	red	1846	Lord Barrington (9308)	do. 54th.
Do. 59th	roan	1847	2nd Duke of Oxford (9046)	do. 56th.
Do. 60th	red	1847	4th Duke of Northumberland	do. 54th.
Do. 61st	red roan	1848	2nd Duke of Oxford	do. 51st.
Do. 62nd	red & white	1848	do.	do. 56th.
Do. 63rd	roan	1848	do.	do. 54th.
Do. 64th	red	1849	do.	do. 55th.

A well-known writer* says:—

"There are few short-horn breeders who have not studied the Herd Book more or less, but though that source of information on the important subject of pedigrees is as perfect as it well can be expected to be, it will afford to the general investigator very little

---

* Dunelmensis. *Mark Lane Express*, 19th April, 1858.

assistance as a practical breeder, and he must apply to other sources for the information he wants, and which is essential to his success. The fact is that certain tribes of short-horns, or families, have established characters as good breeders, which has raised them in public estimation, and made them valuable; and, though, in an old established family a strong cross may sometimes answer, yet, the case in which it does is the exception, and no short-horn breeder would attempt it, if he could possibly avoid the risk. Now, with regard to the instance which you produce of Mr. Bates' 'Cleveland Lad,' it would be a dangerous precedent, for two reasons, first, the Matchem cow was not almost without a pedigree. She was bred by Mr. Brown. Now Mr. Brown, though he kept no written pedigrees, was a breeder of shorn-horns for, I believe, half a century. He bred from Mr. Mason's best bulls, of which Matchem was one. This was well-known to Mr. Bates, and when the cow produced a first-rate bull, Mr. Bates used him with confidence, well knowing the purity of the blood. It is well-known to those persons who were acquainted with Mr. Bates, that he did not speak very favourably of other short-horn herds, and that he very frequently disparaged the Mason blood. He therefore did not search after the pedigree of the Matchem cow, because he knew well it would lead him up to Mason; he preferred leaving the pedigree in the state we have it in the Herd Book; however, there is more than one short-horn breeder now alive who has heard Mr. Bates describe the Matchem cow and say, she had no inconsiderable dash of Mason blood. Mr. Mason's sale of short-horns was in 1829. Two Irish gentlemen attended and were purchasers, and after the sale was over, this Mr. Brown, the breeder of the Matchem cow, came up to them, seeing they wanted short-horns, and told them then he had a very good, well bred herd, that they were always crossed by Mr. Mason's best bulls, and that if they went home with him he would sell them as good as any sold at Mr. Mason's sale, and for half the money paid there. They did go to Mr. Brown's, and did purchase from him, and there can be no doubt his herd was purely bred for many generations. Matchem's dam was by Farmer, and Farmer was out of a cow entirely unknown. Now this would lead many to believe that Farmer was not well bred,

v

and I rather think the supposition would be incorrect, for the dam of Farmer was bred by Mr. John Newby, a man who bred good short-horns of pure blood, though he did not keep his herd book."

I will now trace the descendants of the Matchem cow, Herd Book, Volume III.,* bred by Mr. Brown and bought by Mr. Bates.

	Colour.	When Calved.	Sire.	Dam.
Oxford Premium Cow......	...roan.........	1834	Duke of Cleveland (1937)............	Matchem Cow.
Do. 2nd...	do.	...1839...	Short Tail (2621).....	do.
Do. 3rd	...white.........	1840...	Dk. of Northmd.	...Ox. Prem. Cow.
Do. 4th	...red & white..	1843...	do.	... do.
Do. 5th	...roan.........	1844...	do.	...Oxford 2nd.
Do. 6th	...red	1846...	2nd Dk. of Nrthd...	do.
Do. 7th	...roan	1847...	4th Dk. of Nrthd...	do.
Do. 8th ...	do.	...1847...	2nd Cleveland Lad.	Oxford 5th.
Do. 9th ...	do.	...1848...	3rd Dk. of York.....	do. 2nd.
Do. 10th	...red & white..	1848...	do.	... do. 5th.
Do. 11th	...dark roan .	1849...	4th Dk. of York.....	do. 6th.
Do. 12th	...light roan...	1849...	do.	... do. 4th.
Do. 13th	...roan.........	1850...	3rd Dk. of York.....	do. 5th.
Do. 14th ...	do.	1850...	do.	... do. 2nd.

The Waterloo tribe (the first calved in 1829) was descended from a cow bought in 1831, at Thorpe, in the County of Durham, and had two crosses by Waterloo (2816). The following is in the handwriting of Mr. Bates:—

"I have seen the gentleman who bred the Waterloo cow lately, and he stated to me that he and his father had had the breed for fifty years, and that they were well descended all that time, having

---

* Matchem (2281) cow, white, calved in 1827, bred by Mr. Brown, the property of Mr. Bates, her dam by Young Wynyard (2850) sometimes called Young Wellington.

had a son of Comet, and other blood before the cross with Waterloo bull (2816). I purchased the Waterloo cow in 1831."

		Colour.	When Calved.	Sire.	Dam.	
Waterloo	2nd	roan	1832	Belvedere	Waterloo cow.	
Do.	3rd	red	1835	Norfolk	do.	
Do.	4th	red & white	1840	Cleveland Lad	Waterloo 3rd.	
Do.	5th	roan	1841	Duke of Northumberland	do.	
Do.	6th	red	1842	do.	do.	
Do.	7th	red & white	1843	do.	do.	4th.
Do.	8th	do.	1844	2nd Cleveland Lad	do.	6th.
Do.	9th	red roan	1847	do.	do.	
Do.	10th	red		4th Dk. of Northumberland	do.	8th.
Do.	11th	yellow red, and white	1848	2nd Duke of Oxford	do.	4th.
Do.	12th	red	1849	3rd Dk. of York	do.	
Do.	13th	roan		3rd Dk. of Oxford (9047)	do.	9th.

The first of the Red Rose or (after the Cambridge Meeting called) Cambridge Rose tribe, are as follow. Red calved in 1811.*

		Colour.	When Calved.	Sire.	Dam.	
Red Rose	2nd	red & white	1821	His Grace	Red Rose 1st.	
Do.	3rd	do.	1824	The Earl	do.	2nd.
Do.	4th	do.	1824	The 2nd Earl	do.	
Do.	5th	do.	1827	2nd Hubback	do.	
Do.	6th	do.		do.	do.	3rd.
Do.	7th	do.	1828	do.	do.	2nd.
Do.	8th	do.	1829	do.	do.	
Do.	9th	do.	1831	do.	do.	
Do.	10th	do.	1832	Bertram	do.	7th.

* Bred by Mr. Hustler, property of Mr. T. Bates; by Yarborough, dam (bred by Mr. R. Colling, and called the American cow), by Favourite, 2nd dam by Punch, 3rd dam by Foljamb, 4th dam by Hubback.

	Colour.	When Calved.	Sire.	Dam.
Red Rose 11th or Rose of Sharon	roan	....	...Belvedere	...Red Rose 5th.
Red Rose 12th	red & white	1833	do.	do. 8th.
Red Rose 13th or Cambridge Premium Rose	roan	1834	do.	do. 9th.
Cambridge Rose 2nd	do.	1837	do.	Cambridge Prem. Rose
Do. 3rd	white	1844	Dk. of Northumberland	Cam. Rose 2nd
Do. 4th	red & white		2nd Cleveland Lad	Cambridge Prem. Rose
Do. 5th	roan	1846	do.	Cam. Rose 2nd
Do. 6th	do.	1848	3rd Duke of York	do. 5th.
Do. 7th	red	1849	do.	do.

"At Mr. Robert Colling's sale, at Barmpton, in 1818, the Red Rose tribe of cattle brought the highest prices. A bull sold for 621 guineas. Mr. Arbuthnot paid 320 guineas for one cow of this tribe, and Lord Althorp 300 guineas for another, and many more at high prices. But the one at 300 *guineas*, when going with the Acklam *Red Rose* at Halton Castle, in Northumberland, was very inferior to the latter in 1820, although they were out of own sisters."—*Note by Mr. Bates.*

The Wildeyes tribe (originally Wildair) was descended from a heifer calf bought at Mr. Parrington's sale in 1831, at Middlesborough farm, and were descended from Dobinson's stock, purchased in Holland.

Mr. Parrington occupied the farm of Middlesborough, on the south bank of the Tees, below Stockton, and in 1831, the farm house and a few cottages were the only buildings on the land on which the town of Middlesborough now stands, and the centre of the iron trade of the world. A town with a corporation, a member of Parliament, and upwards of 60,000 inhabitants. Mr. Parrington was a well known breeder of short-horns. His stock were sold in 1831. I attended the sale with Mr. Bates and bid for the lots which he bought. The calf which he kept was called Wildair, but when she came to Kirklevington somehow she got the name of Wildeyes. The pedigree is as follows:—

WILDEYES by Emperor (1975).
    1st dam by Wonderful (700).
    2nd dam by Cleveland (145).
    3rd dam by Butterfly (104).
    4th dam by Hollon's bull (313).
    5th dam by Mowbray's bull (2342).
    6th dam by Masterman's bull (422),—descended from Michael Dobinson's stock.

The estimation in which this tribe were held by Mr. Bates appears in the above pages. He always spoke of it as giving him the (Dobinson) only good blood the Colling herds did not contain. His uncle, William Bates's herd, at Chollerton, in North Tyne, were long celebrated as excellent, and for many years the steers from it were exhibited in Newcastle as prize beef. William Bates bred much in conjunction with the Messrs. Culley. He was well known for his mathematical knowledge, and was a Commissioner for the Enclosure of nearly all the Waste and Common Lands in the North of England. The great Inglewood Forest in Cumberland, which extended from near Carlisle to Penrith, was divided and enclosed by him. Mr. Donkin, of Sandhoe, so often mentioned in short-horn history, was brother-in-law of Willliam Bates. He belonged to a talented family. One was a celebrated mechanical engineer in London, and one Professor of Astronomy at Oxford. Sandhoe is opposite Dilston on the Tyne. The stock, cattle and sheep of Mr. Donkin, and his tillage management, were long equal to the best in the North of England. Mr. Bates and Mr. Donkin were often in dispute about breeding. Mr. Donkin did not like the lectures and advice of Mr. Bates on the subject.

I append the descendants of this tribe, bred by Mr. Bates.

	Colour.	When Calved.	Sire.	Dam.	
Wildeyes 2nd	roan	1835	Belvedere	Wildeyes.	
Do. 3rd	do.	1836	do.	do.	
Do. 4th	red	1837	Short Tail	do.	
Do. 5th	roan	1839	do.	do.	
Do. 6th	white	1840	Dk. of Nrthmd	do.	2nd.
Do. 7th	do.	1841	do.	do.	3rd.
Do. 8th	roan	1842	do.	do.	2nd.
Do. 9th	red & white	1842	do.	do.	3rd.
Do. 10th	roan	1843	4th Dk. of Nrthd	do.	2nd.
Do. 11th	red & white	1843	Dk. of Nrthmd	Wildeyes.	
Do. 12th	roan	1844	2nd Dk. of Nrthd.	do.	5th.
Do. 13th	do.	1844	4th Dk. of Nrthd.	do.	7th.
Do. 14th	red & white	1845	Dk. of Nrthd	do.	3rd.
Do. 15th	do.	1845	4th Dk. of Nrthd.	do.	8th.
Do. 16th	roan	1845	2nd Dk. of Nrthd.	Wildeyes.	
Do. 17th	red & white	1845	do.	do.	5th.
Do. 18th	roan	1845	2nd Clvlnd Lad	do.	3rd.
Do. 19th	do.	1846	2nd Dk. of Oxford	do.	10th.
Do. 20th	red & white	1846	Fl. of Lvrpl (9061)	do.	9th.
Do. 21st	roan	1847	2nd Clvlnd Lad	do.	10th.
Do. 22nd	do.	1847	do.	do.	8th.
Do. 23rd	do.	1847	do.	do.	9th.

	Colour.	When Calved.	Sire.	Dam.
Wildeyes 24th	roan	1847	2nd Clvlnd Lad	Wildeyes 5th.
Do. 25th	yellow red, & white	1848	do.	do. 12th.
Do. 26th	red	1848	do.	do. 5th.
Do. 27th	roan	1848	do.	do. 17th.
Do. 28th	do.	1849	do.	do. 16th.
Do. 29th	light roan	1849	3rd Dk. of York	do. 19th.
Do. 30th	white	1849	3rd D. of Ox. (9047)	do. 7th

The first cow of the Foggathorpe tribe was bought, as related by Mr. Bates, of Mr. Edwards, of Market Weighton. The following is her pedigree:—

FOGGATHORPE, roan, calved in 1830, by Malbro' (1189).

    1st dam Rosebud, by Ebor (997).

    2nd dam Tulip, by Regent (546).

    3rd dam Primrose, by North Star (459).

    4th dam by Mr. R. Colling's White bull (151).

    5th dam bred by Mr. R. Colling.

The following is a note by Mr. Bates:—

"Mr. Seaton did not keep an account of the pedigrees of the cattle he bought of Mr. Robt. Colling, but agreed to give Mr. Robt. Colling a hundred guineas each for whatever cows he pleased to part with for many years together. Mr. Seaton used Mr. Robt. Colling's White bull for many years together, and Foggathorpe's character resembled that of Princess, bought by the late Sir H. V. Tempest, and might have been of the same tribe as Princess, but of this there is no account now to be had."

This cow was 10 years old when bought by Mr. Bates and not expected to breed. She was second prize in the show when the Cambridge Rose was rejected. Mr. Bates always said she was very superior to the cow that got the first prize, but both very inferior to the Cambridge cow. I well remember her coming to Kirklevington, and somehow Mr. Bates always spoke very highly of this tribe. They had some cross that he thought very highly of. Mr. Bates had this cow tied up next the rejected Cambridge cow. Mr. Bates took me to see her and asked my opinion of his purchase. I looked at her and expressed a very favourable opinion of her shape and points. He then told me to put my hands upon her and then upon the Cambridge cow, I immediately felt the difference, and Mr. Bates ever after took his visitors to see the two cows, and to test them by the touch. The two cows were kept together and fed the same and sold together. I do not remember his saying so, but certainly we all got the impression that Mr. Bates had bought this cow to show visitors how ignorant judges at the great shows were of what constituted the chief value of improved short-horns. There was no doubt as to the superiority of the Cambridge cow over this cow, although a good one. The following were the females bred at Kirklevington. The produce certainly showed the value of his opinion that all breeds would improve by his own blood.

	Colour.	When Calved.	Sire.	Dam.
Foggathorpe 2nd...	white	1840...	Duke of Northumberland.	Foggathorpe.
Foggathorpe 3rd..	red & white...	1841...	do.	do.
Foggathorpe 4th...	light roan	1842...	do.	do.
Foggathorpe 5th..	roan	1846...	4th D. of Nrthd...	do. 2nd.
Foggathorpe 6th...	do.	1850...	3rd. D. of Oxford.	do. 4th.

PART XV.

Intended to retire in 1851—Mr. Pippet and Appreciation of Stock—Fortune—Opinions on Stock—Ingratitude and Anonymous Slanders—American Appreciation—Humbug—Father and Mother, Deaths, Relations, and Old Reminiscences—Engaged to be Married—Consumption—Advice of Scotch Legal Judge—Obituary notice in *Farmer's Magazine*—Death—Monuments.

Mr. Bates had long contemplated selling off his herd and retiring altogether from all active business, in the year 1851, and always expressed great anxiety that they should go into the hands of some one who would carry out his ideas as to the crossing and breeding of them. I am here reminded by Wm. Pippet, Esq., of Coughton, of Mr Bates's words to him which he has never forgot, viz., "I shall never live to see the day, you may, when my stock will be duly appreciated." Mr. Bates had written to Mr. Pippet in December, 1847, as follows:—

"In answer to your letter of the 6th inst., enquiring the breeding of the last bull Mr. Glass had from here, as you have bought the stock by him. He was by Duke of Northumberland (1940) dam by Short Tail (2621) second dam by Belvedere (1706) third dam by son of second Hubback (2683) being four crosses of blood that cannot be exceeded and leaving only one-sixteenth of the original cow, which was a very good one bought in at the market, the seller giving a long pedigree but not of my blood. (He then mentioned Coates and the Herd Book). I may give an authentic history of the short-horns when those who have my blood will receive the benefit therefrom."

Mr. Bates often said that money-making was not his object in perfecting the race of short-horns. If that had been his motive and he had invested his fortune in the funds, he would have been a very wealthy man. He had begun the world when only over age, with a fortune in round figures of £20,000. His personal expenditure was trifling, but from charitable and religious expenditure, his fortune did not much increase. He could afford to devote his income to his favourite herd, otherwise it would probably have been dispersed and long forgotten.

I have intentionally avoided noticing any of the controversies which Mr. Bates had respecting stock. He certainly had no idea of cooking pedigrees or making things in that line pleasant. He was not prone to abuse stock except when it was brought before the public, and he then considered it a public duty to use his knowledge and give his opinion. He was often asked to visit herds and did so, but he never praised, even to please an old friend or kind host. On one occasion he paid a visit to an old friend who had a fine herd and who expected Mr. Bates to praise them; he, however, was silent, and when pressed for an opinion he said they were very fine cattle but not short-horns. They were kyloes. His friend then explained that they had been bred from a kyloe cow. On another occasion, when in like manner pressed for an opinion, they met his friend's boys and the nursemaid; Mr. Bates then exclaimed, I now see some stock I can admire. Mr. Bates was always very fond of, and attentive to children.

My attention has been called to an announcement of the sale of the Farnley Hall short-horns in the newspapers, and I have only time to notice it shortly. It states:—"The Farnley Hall catalogue of the sale, which Mr. Strafford will conduct at Farnley Hall on the 1st of August will be read with interest by short-horn breeders up and down the country." It then describes the various animals, and they will be sold where Mr. Strafford has sold so many splended cattle, and often giving names and pedigrees. "Reformer, from a cow perseveringly traduced by the much calumniating Thomas Bates." Probably this is the production of an individual who is indebted to Mr. Bates for every thing he possesses, and who, in return for kindness, confidence, and hospitality, became the means or the vehicle of making anonymous attacks on Mr. Bates and his herd. I have not time now to peruse all the correspondence on the subject of this cow. Mr. Bates, I believe, warned the American gentlemen against produce from her.* Mr. Bates had long ago taken a gentleman to inspect Mr. Baker's herd, where he often visited and found them very much mixed in blood, and advised his friend not to buy them. Mr. Bates had, after Lord Althorp's purchases, obtained many well bred animals for this friend. Mr. Bates afterwards found that he had purchased from Mr. Baker's herd, and in answer to remonstrance said he had a good connexion in trade to America for cattle which he must keep up, and the Americans knew nothing of breeding, they only wanted long pedigrees. A bull descended from this cow had been sold for £14, and then for £16, and very soon after was sent to

---

* I have just time to state other facts respecting this cow. We have seen what Mr. Bates wrote about crossing his stock with the blood of this cow, and the probable results. This was all verified. I well recollect that Mr. Bates received a letter from America, stating that the stock had turned out badly as he had predicted, and asking him to send the Ohio Company any stock he could spare, and that they left that and the price he would charge entirely to himself. Mr. Bates tore up the letter, and sent no stock. Only nine animals from his own herd went to America in his lifetime.

America at nearly 200 guineas. The cow was purchased at Yarm fair and turned out a very good one, and Mr. Bates was requested to trace her breeding, which he did, and found she was from a blue-black cow with no pedigree, and of a colour and description that every one said no short-horn ever had. Mr. Bates would not conceal Clarke's bull nor his knowledge of this cow when requisite, but he made no anonymous announcements. I hope the next catalogue will be altered to " the cow openly exposed by the truth-speaking and HUMBUG denouncing Thomas Bates."

Mr. George Bates, his father, died in 1816, aged 84. No man ever was more respected or regretted. I recently visited my maternal uncle, now aged above 90, with all his faculties, except sight, quite perfect. He was loud in his praises of the farms and stock of Mr. George Bates, and the good he did in the country, and regretted that there were few (if any) such men left in the country. Mr. George Bates, for many years, had not a pauper in the township which he occupied, and many of the persons employed by him became bailiffs and stewards, or got into business for themselves. They were all elevated by the example he set them. Mr. G. Bates tried almost every new implement or crop, but he used work oxen to the last, and splendid animals they became.

Mr. Bates often expressed his opinion that he would live to the age of his mother, who died in 1822, aged 92 years. I well recollect the " old lady," and the deference always paid to her. When past 80, she rode a white pony, and afterwards used a pillion, perhaps the last seen in the county. Mr. Bates always spoke with great affection of her. He was her favourite child. She had only visited her native county once after her marriage, which was when she took her sons to visit her cousins at Gregygnog. But she and her sister Joyous Moore, who resided and died with her, kept up a large correspondence with their relatives and friends in Wales and the Borders. It is now, perhaps, amusing to read the accounts of the elections, races, county gatherings, and balls, in the last century, and there may be some who would be interested in the announcements and expected announcements of conquests made by the young ladies. But, the style of the letters changed, and there came accounts of loss of many relatives and friends, by battle or disease, in the service of their country, and a gradual disappearance, and announcements of deaths and funerals, often at The More, until, of the first cousins who once numbered above thirty, Mrs. Bates was the last survivor. She was not fortunate in succeeding to the goods of her relatives. Mr. Bates always set great value on his handsome silver tea service, which was the bequest of Mrs. Catherine More, of Linley, to her god-daughter Joyous More. The only descendants of the cousins were the Edwardes family. Sir Herbert Edwardes, the distinguished Indian officer, being one, and, I believe, they are now represented by the Hope-Edwardes family in Shropshire. Mrs. Bates, like all old people, had a most lively recollection of the olden times, and talked of the widows of her uncles killed at the Battle of the Boyne, and of her recollections of relatives who had taken active parts in the Irish Wars and Rebellions of 1715 and 1748. Mrs. Bates always considered that if her son had been called Arthur Blayney, he would have got the family estates.

Mr. Bates was under the middle size, but with a strongly built frame. He never married. I have heard many romantic and curious stories of his loves and disappointments, but I believe without the slightest foundation, and that in fact he never even spoke to the parties. He was however engaged, as appears from his mother and aunt's letters to their relatives, to a young lady in every way a suitable and desirable match, but she died of consumption, and I believe he never got over his loss, and that when he spoke of hereditary disease and consumption in cattle, his feelings were much moved and painful recollections brought to his mind. His friends often urged him to change his state. He was always most attentive to ladies, and in his later years he was often urged to look speedily after some eligible widows. Baron Hepburn, one of the judges of the Court of Exchequer in Scotland, was an intimate friend of Mr. Bates, and also of Mr. Curwen, and Mr. Rennie, and the leading agriculturists in the East Lothians. I venture to insert the letter of advice to Mr. Bates on matrimony and agriculture, by this spirited Tory Scotch Judge.

"Smeaton, Prestonkirk, 7th July, 1813.

" My dear Bates,—Some eight or ten days ago I had the pleasure to receive your obliging favour of the 17th ult. while in Edinburgh attending my official duty in Exchequer, which concluded on Monday last, and I came here yesterday. I rejoice to hear such flattering accounts of your young stock. It always affords me real pleasure to see genius and industry rewarded. I always told you that the cross of the Scotch breed would answer your expectation, and you may recollect I advised you to try it on yourself, and now that you have proved it so successfully in one species of stock I trust you will listen to my advice and carry it one step further. Do come down to us when you can spare as much time and cast your eye over our fair ones, and to the lass you like best, sing our Scotch song, which begins thus—

'Tell me Lassie, will you marry me now,
For I canna come every day to woo.'

I have sent your compliments to my neighbours, Messrs. Rennie and Brown, but I delayed writing until I could tell you of my farm, and I have the pleasure to say that in every particular it would do your heart good to see how promising it really is. My autumn sowed wheats are well planted, and have been now about a week in bloom, so I think they will cut by the second week of next month. My oats are strong and showing the ear. My peas and beans completely cover the ground—a vigorous growth and full of bloom, and setting remarkable well. My Roota Baga is far advanced, and my turnips generally in the rough leaf, and both fully planted. My hay harvest somewhat advanced. Plenty of clover plants, but late and soft. I expect a great second crop, and I have plenty of old straw to mix up with it. My practice is to carry the straw to the field and to mix it with the clover when cut. This is a little more expensive, as I must carry the straw out and home again; but the new and the old incorporate so intimately that my horses are fonder of it than of pure hay. 20 head of cattle paid me of nett return £25 a head for their keep, and my sheep netted £1 19s. a head. Since 1798 I have not had so good a crop of all corn as 1812 has yielded, and the prices last year nearly double. What I say of my own, for men generally are fond of speaking of themselves, I may truly say of all my neighbours. So we are contented, quiet and happy, and I thank God we have no rascally drunken discontented weavers amongst us, 'our pleasures to 'alloy.'' Our staples are, wheat and bairns. I wish you would begin to raise recruits the matrimonial way, and give the king some good healthful subjects. Try the cross and fill up the waste of war. Your butter last year was super-excellent, and my wife is your humble suitor this year for a single kit if you can spare it. I congratulate and rejoice with you on the glorious news from the Peninsula, and I by no means regret that the Catholic question was lost. Adieu, and believe me always with esteem, sincerely yours.

"GEO. BUCHAN KEPBURN."

The following memoir of Mr. Bates appeared in the *Farmers' Magazine*, in January, 1850. Speaking of Mr. Bates, the writer says:—

"As a man, there were few who enjoyed a wider range of popularity than Mr. Bates. The employment he gave to the poor did not more ingratiate him in their favour, than the unvarying and unmingled kindness he at all times displayed, whether in providing for their cheap and comfortable shelter in his cottages, or ministering to their wants in sickness, infirmity, or age. His kindness as a neighbour was beyond all praise. Scarcely one of the farmers, whose cold barren clay farms surrounded him, but could bear witness to some act of disinterested sympathy; and a stranger would have witnessed with surprise the influence his name and his opinion had upon them; while his word would be more relied upon than many men's bond. In hospitality to all comers he was seldom equalled; his house was open to every one of whatever grade, from the Peer or Member of Parliament, down to the small undistinguished farmer, and the longer the visitor stayed, and the more he partook of his liberality, the more welcome he was. On one occasion, two very celebrated short-horn breeders intimated a wish to spend a day at Kirklevington and examine his stock; he immediately wrote in reply, that it was impossible for them to examine them thoroughly in one day, and that they must make up their minds to spend three with him, in fact his house was the home of all who entered it. They had a welcome truly English; no pains nor unostentatious attention were spared to make them comfortable and happy; while his long stories, founded on bye-gone experience, of great breeders of the early part of his life, the dark ages of short-horns, were so amusing, that the time flew in his society completely unawares, and to no one did it do so more swiftly than himself. In fact but once set him on with his anecdotes, and there never seemed to be a termination, nor even a breaking place when the meeting could be separated. However inconsistent with all this power it may appear, he was often in hot water with some of his opulent and influential neighbours, and has more than once driven the bench of magistrates from court to court at enormous expence. Amongst many of these he was looked upon as meddling, overbearing, and litigious; certainly the pertinacity with which he opposed many of their measures of a public nature, and the very great expenses he incurred, might seem to justify the opinion. But those who entertained this opinion did not understand Mr. Bates. The dispute was not private nor personal; it was of a purely public origin, and with him the course he took was looked upon as a great public duty, and one that he felt he would be false to himself to abandon. His litigiousness, therefore, was but a nice and discriminating view of public duty; and had the magistracy used a little conciliation due to a man so well disposed and time-honoured, they might have achieved any concession they required, consistent with great public responsibility, which Mr. Bates held with most scrupulous conscientiousness. On this point, perhaps, his judgment did not always equal his zeal and perceptions of right. Convince his judgment or appeal to his feelings, and he was gentle and yielding; but once rouse his opposition and he was as untiring in his warfare as he was staunch and unflinching in his character. As one instance of the general benefit arising from his exertions, we may name his procuring a clause to be inserted in the Highway Act, removing the power of electing surveyors out of the hands of the magistracy, and placing it in those of the rate-payers.

"Mr. Bates was fond of public life, and was not altogether free from a love of excitement. Once he risked the cost and labour of setting afloat a county contest for the representation; and had indeed great delight in addressing the public, using very strong language, and always appearing in earnest. He wrote a vast number of letters to the newspapers, mainly on politics of agriculture, and was always at his post at a county meeting, or election, where anything agricultural was the subject of investigation or remark. His writing, though not elegant nor classic, was true and forcible, and he had a remarkable tact in making facts bear upon his propositions, as well as a wonderful readiness in calculation and mental arithmetic. It was, however, Mr. Bates's character as a christian which gained him the large amount of respect he so generally secured; and an undeviating course of moral conduct, absolutely untainted and unimpeachable, gave him a standing, which, though it might for the moment excite the ridicule of the thoughtless, generally created a

W

real respect in their minds. At a period when a profession of religion, was by no means so fashionable as it is at present, he would dare ridicule and scorn.

"His zeal and liberality went hand-in-hand. On one occasion he heard that a living, in a parish in which he felt concerned, was about to be given to a clergyman whom he believed to be unworthy; and though the kindest man living, he was determined to prevent it. He wrote to the bishop and the archdeacon, confronted the parties interested, and, by dint of persevering opposition, effected the rescinding of the gift, and obtained an appointment congenial to his wishes. Active in mind, temperate in his habits, nay, I may say abstemious, for he tasted no intoxicating liquors for some years before his death; and living almost in the open-air, he knew little of disease, and seldom, if ever, consulted a surgeon. A month before death, however, his health began to fail, a disease of the kidneys became painful and harassing, and he went to Redcar to try the effect of the sea-air, but which, so far from removing, seemed only to increase the malady. It was some time before he could be prevailed on to consult a medical adviser, and when he did, he refused the greater part of the medicine."

He gradually sank and died on the 25th of July, 1849, aged 75 years, and was buried in the churchyard at Kirklevington. It is intended by his nephew to insert a memorial window in the now restored church to his memory. A monument was erected to his memory by a few of his friends and admirers of his exertions in the cause of stock breeding, with the following inscription:—

THIS MEMORIAL
OF
THOMAS BATES,
OF KIRKLEVINGTON,
ONE OF THE MOST DISTINGUISHED BREEDERS OF
SHORT-HORN CATTLE,
IS RAISED BY A FEW FRIENDS WHO APPRECIATE
HIS LABOURS FOR THE IMPROVEMENT OF
BRITISH STOCK,
AND RESPECT HIS CHARACTER.
BORN 21ST JUNE, 1776,
DIED 26TH JULY, 1849.

PART XVI.

Kirklevington Sale—Particulars—Notice in *Farmers' Magazine*—Why the Prices go Low—Comments on Lord Ducie and Jonas Webb—Racing Phrase—Lord Ducie's anxiety in Breeding—3rd and 5th Duke of York—Mr. Tanqueray—Tortworth Sale and the Americans—Col. Gunter—Marquis of Exeter—Lord Feversham, his judgment and the Bates's Blood—Willis's Rooms Sale—Importation from America—*Punch* on the Windsor Sale—History of Divers Animals—Grand Duchesses, &c.

Mr. Edward Bates was the only nephew of Mr. Bates engaged in agriculture, and was settled in Germany, and had no time or opportunity of attending to a herd, so that it came to be sold on the 9th of May, 1850, and I subjoin the results:—

SYNOPSIS OF THE PEDIGREE, PRICES, AND PURCHASERS OF THE KIRKLEVINGTON HERD.

COWS, HEIFERS, AND HEIFER CALVES.

DUCHESS FAMILY.

Name.	Bulled by	Price. £ s.	By whom purchased.
Duchess 51st	2nd Duke of Oxford, April 1	63 0	Mr. S. E. Bolden, Lancaster.
Duchess 54th	Ditto, Nov. 8	94 10	Mr. Eastwood, Burnley.
Duchess 55th	Ditto, April 1	110 5	Earl Ducie.
Duchess 56th	Grand Duke, April 5	54 12	Mr. Ambler.
Duchess 59th	Ditto, Nov. 23	210 0	Earl Ducie.
Duchess 61st	——	105 0	Lord Feversham.
Duchess 62nd	——	126 0	Mr. Champion.
Duchess 64th	——	162 15	Earl Ducie.

OXFORD FAMILY.

Name.	Bulled by	Price.	By whom purchased.
Oxford 2nd	Duke of Richmond, May 5	54 12	Marquis of Exeter.

Name.	Bulled by	Price. £ s.	By whom purchased.
Oxford 4th	3rd Duke of York, Nov. 2	28 7	Mr. E. James.
Oxford 5th	Ditto, April 27	74 11	Mr. Morris.
Oxford 6th	Grand Duke, Feb. 6	131 5	Earl Ducie.
Oxford 9th		42 0	Mr. A. Maynard.
Oxford 10th		53 11	Mr. Morris.
Oxford 11th		131 5	Earl Ducie.
Oxford 12th		85 1	Lord Feversham.
Oxford 13th		66 3	Mr. Becar, of the U. S., America.
Oxford 14th		21 0	Mr. Downes.

WATERLOO FAMILY.

Name.	Bulled by	Price.	By whom purchased.
Waterloo 4th	3rd Duke of Oxford, April 26	22 1	Mr. Singleton.
Waterloo 9th	Ditto, Nov. 9	79 16	Mr. Ashton.
Waterloo 10th	Grand Duke, April 25	63 0	Mr. A. Maynard.
Waterloo 11th	4th Dk. of York, Feb. 26	73 10	Mr. Eastwood.
Waterloo 12th		44 2	Mr. Cruickshank.
Waterloo 13th		74 11	Mr. Hay Shethin.

CAMBRIDGE ROSE FAMILY.

Name.	Bulled by	Price.	By whom purchased.
Cam. Rose 5th	Not bulled	47 5	Mr. S. E. Bolden.
Cam. Rose 6th		73 10	Mr. H. Coombe.
Cam. Rose 7th		26 5	Mr. Downes.

WILDEYES FAMILY.

Name.	Bulled by	Price.	By whom purchased.
Wildeyes 5th	3rd Duke of Oxford, April 21	21 0	Mr. A. Stevens.
Wildeyes 7th	Grand Duke, March 1	24 3	Mr. Jefferson.
Wildeyes 8th	Ditto, Jan. 15	42 0	Marquis of Exeter.
Wildeyes 14th	Duke of Richmond, May 8	30 9	Mr. Jonas Webb.
Wildeyes 15th	2nd Duke of Oxford, March 24	32 11	Mr. Featherstonhaugh.
Wildeyes 16th	3rd Duke of York, April 18	23 2	Mr. Higgs.
Wildeyes 17th	4th Duke of York, Jan. 20	43 1	Mr. Faviell.
Wildeyes 19th	Grand Duke, Nov. 26	63 0	Mr. Cartwright.

Name.	Bulled by	Price.	By whom purchased.
		£ s.	
Wildeyes 21st...4th	Duke of York, April 4th	49 7	Mr. A. Morrison.
Wildeyes 22nd..3rd	Ditto, Jan. 24	105 0	Mr. Champion.
Wildeyes 23rd...3rd	Duke of York, Feb. 26	105 0	Mr. A. Maynard.
Wildeyes 24th...	Ditto, Jan. 12	42 0	Mr. Drummond.
Wildeyes 25th...4th	Duke of York, Nov. 23	74 11	Mr. B. Baxter.
Wildeyes 26th...		31 10	Mr. Haigh.
Wildeyes 27th...		45 3	Mr. N. Cartwright
Wildeyes 28th...		27 6	Mr. E. Bates.
Wildeyes 29th...		39 18	Lord Feversham.
Wildeyes 30th...		24 3	Mr. Townshend.

FOGGATHORPE FAMILY.

Name.	Bulled by	Price.	By whom purchased.
Foggathorpe 2nd.	Grand Duke, April 18	22 1	Mr. Parker.
Foggathorpe 4th.3rd	Duke of Oxford, April 6	52 10	Mr. Sanday.
Foggathorpe 6th.		31 10	Mr. Gardiner.

BULLS AND BULL CALVES.

DUCHESS FAMILY.

Name.	Colour.	Calved.	Sire.	Dam.	Price.	By whom purchased.
Duke of Richmond (7996)	roan	Aug., 1844	2nd Cleveland Lad (3408)	Duchess 50th	£126 0 0	Mr. A. L. Maynard.
Third Duke of York (10166)	red	Oct., 1845	4th Duke of Northumberland (3649)	Duchess 51st	74 11 0	Mr. G. D. Trotter.
Fourth Duke of York (10167)	roan	Dec., 1846	2nd Duke of Oxford (9046)	ditto.	210 0 0	Earl Ducie.
Grand Duke (10284)	red	Feb., 1848	2nd Cleveland Lad (3408)	Duchess 55th	215 5 0	Mr. Hay.
Duke of Athol (10150)	red	Sept., 1849	2nd Dk. of Oxford (9046)	Duchess 54th	42 0 0	Mr. Parker.
Fifth Duke of York (10168)	white	Oct., 1849	ditto.	Duchess 51st	33 12 0	Mr. R. Bell.

OXFORD FAMILY.

Name.	Colour.	Calved.	Sire.	Dam.	Price.	By whom purchased.
Second Duke of Oxford (9046)	roan	Aug., 1843	Duke of Northumberland	Oxford 2nd	110 5 0	Earl Howe.
Third Duke of Oxford (9047)	roan	Oct., 1845	2nd Duke of Northumberland	Oxford 2nd	64 1 0	Mr. Robinson.
Beverley (9964)	red & w.	Oct., 1848	2nd Earl of Beverley (5963)	Oxford 4th	32 11 0	Mr. Townshend.

BULLS AND BULL CALVES *Continued.*

WILDEYES FAMILY.

Name.	Colour.	Calved.	Sire.	Dam.	Price.	By whom purchased.
Lord George Bentinck (9317)	roan	April, 1845	2nd Duke of Northumberland	Wildeyes 2nd	£29 8 0	Mr. Annett.
Parrington (10590)	red & w.	Dec., 1847	2nd Cleveland Lad	Wildeyes 15th	25 4 0	Mr. Fisher.
Red Rover (10692)	red & w.	Sept., 1848	ditto.	Wildeyes 8th	36 15 0	Mr. E. Bates.
Balco (9918)	red	Feb., 1849	4th Duke of York	Wildeyes 15th	162 15 0	Earl of Burlington.
Retriever (10707)	light roan	Aug., 1849	3rd Dk. of Oxford	Wildeyes 8th	52 10 0	Earl of Carlisle.
Crusader	white	Jan., 1850	2nd Dk. of Oxford	Wildeyes 21st	42 0 0	Mr. Blackstock.
Wonderful	red & w.	Jan., 1850	ditto.	Wildeyes 15th	31 10 0	Mr. H. Smith.

FOGGATHORPE FAMILY.

Name.	Colour.	Calved.	Sire.	Dam.	Price.	By whom purchased.
Euclid (9097)	roan	Dec., 1843	2nd Cleveland Lad	Foggathorpe 4th	42 0 0	Duke of Sutherland.
Chevalier (10050)	roan	Aug., 1847	ditto.	Foggathorpe 2nd	43 1 0	Mr. Pullen.
Chieftain (10040)	roan	Aug., 1848	ditto.	ditto.	43 1 0	Mr. Wharton.
Ebor (10184)	light roan	Jan., 1849	3rd Duke of York	Foggathorpe 4th	94 10 0	Lord Feversham.

SUMMARY OF THE SALE OF THE KIRKLEVINGTON HERD OF SHORT-HORN CATTLE.

Families.	No.	Cows.			No.	Heifers.			No.	Heifer Calves.			No.	Bulls.			No.	Bull Calves.			No.	Total.			Average per Head.		
		£	s.	d.		£	s.	d.		£	s.	d.		£	s.	d.		£	s.	d.		£	s.	d.	£	s.	d.
Duchess ...	4	332	7	0	3	441	0	0	1	162	15	0	4	625	16	0	2	75	12	0	14	1627	10	0	116	5	0
Oxford	4	288	15	0	2	95	11	0	4	303	9	0	3	206	17	0	0	......			13	894	12	0	68	16	4
Waterloo ...	2	101	17	0	3	180	12	0	1	74	11	0	0	......			0	......			6	357	0	0	59	10	0
Cambridge...	1	47	5	0	1	73	10	0	1	26	5	0	0	......			0	......			3	147	0	0	49	0	0
Wildeyes ...	9	328	13	0	7	430	10	0	2	64	1	0	4	254	2	0	3	126	0	0	25	1203	6	0	48	2	7
Foggathorpe	2	74	11	0	0	......			1	31	10	0	4	222	12	0	0	......			7	328	13	0	46	19	0
Total	22	1163	8	0	16	1221	3	0	10	662	11	0	15	1309	7	0	5	201	12	0	68	4558	1	0	67	0	7
Average		52	17	7½		76	6	5½		66	5	1		87	5	9¾		40	6	5							

The account of the sale in the *Farmers' Magazine* is so complete and concise that I now copy it.

"The sale of this celebrated herd took place on Thursday, May 9th, 1850, in presence of a company, which, at the lowest estimate, could not be less than five thousand persons, including nearly every breeder of short-horn cattle of note in the United Kingdom, as also breeders from the continent of Europe, and from the United States of America. It may with confidence be maintained that on no similar occasion has so great an interest been excited amongst the breeders of this variety of the ox, so justly the pride of our country, as on that referred to above: and well, indeed, did the herd deserve the far extended fame which attracted such a mighty gathering on the occasion of its dispersion, to be the *nuclei* of new, or to enrich collections already in being, in our sea-girt isles, in Europe, and in the great western quarter of our planet, beyond the Atlantic ocean. To criticise in print a herd whilst it remains the property of the breeder is obviously an improper intermeddling with private property, by which no good purpose can be answered, but which may be productive of controversy, liable to excite vexation. When, however, a herd is dispersed, as on the occasion under consideration, the reason for withholding an opinion of its merits, and of those of the several animals of which it is comprised, ceases: in fact, an event in the annals of rural affairs of such interest and importance as the sale of the Kirklevington herd, not only demands a more permanent record than the ordinary notice in the columns of a newspaper, but now that the cattle in question no longer form a distinct herd, a monument of the incident becomes useful; and no repository for such can be so fitting as the pages of the *Farmers' Magazine*. The herd in question, comprising forty-eight cows, heifers, and heifer calves, and twenty bulls and bull calves, displayed an eminence in every point of excellence which has rarely been attained. In a combination of those qualities which constitute excellence in the short-horn variety of cattle, it may be asserted with confidence, that the Kirklevington herd at the time of its dispersion was unequalled by any other in existence. Magnificent size, straight and broad back, arched and well spread ribs, wide bosom, snug shoulder, clean neck, light feet, small head, prominent and bright but placid eye, were features of usefulness and beauty which distinguished this herd in the very highest degree; whilst the hide is sufficiently thick to indicate an excellent constitution, its elasticity when felt between the fingers and thumb, and its floating under the hand upon the cellular texture beneath, together with the soft and furry texture of the coat, evinced in an extraordinary degree throughout the herd, excellent quality of flesh, and disposition to rapid taking-on fat. In the sixty-eight head of cattle, not one could be characterised as *inferior* or even as *mediocre*—all ranking as first class animals; and when an idea of inferiority arose, it was only in reference to a comparison with some of this splendid herd, which, from their most extraordinary excellence, may demand special notice. The herd consisted of six families :—The Duchess, the Oxford, the Waterloo, the Cambridge Rose, the Wildeyes, and the Foggathorpe, which are here enumerated in succession according to the prices which each realized at the sale; a synopis of the pedigrees, prices, and purchasers, being subjoined, to which it will be sufficient to refer for such particulars. Of the Duchess family, were four cows, three heifers, one heifer-calf, four bulls, and two bull-calves; the first of which that demands especial notice is the Fourth Duke of York. This animal, now the property of Earl Ducie, is the *beau ideal* of bovine excellence.* His magnificent size, and perfection in every point of excellence, entitle him to be considered as the brightest gem of the herd; and if not the very best bull in existence, he certainly cannot be surpassed. Grand Duke, Duchess 54th, and Duchesses 55th, 59th, 61st, 62nd, and 64th, all of the same family, are the finest imaginable specimens of the short-horn tribe. Next in order is the Oxford family, consisting of four cows, two heifers, four heifer-calves, and three bulls, of which Oxford 6th, Oxford 11th, and Second Duke of Oxford, are all animals of extraordinary excellence. The Waterloo and Cambridge Rose families were less numerous than the two preceding. The whole of the animals composing them

* As a proof of this remark and what may be expected from his produce, we beg to observe that the only three calves got by him realized the sum of £379 1s., or £126 7s. each. [Ed. *Far. Mag.*]

possessed great excellence, although inferior to those previously noticed. The Wildeyes, the most extensive family in the herd, consisting of twenty-five head, in which were nine cows, seven heifers, two heifer-calves, four bulls and three bull-calves; and of which Balco, a remarkably fine yearling bull, and two three-year-old heifers, Wildeyes 22nd and 23rd, were prominent lots in the sale. The only remaining family is the Foggathorpe. This family comprised two cows, one heifer-calf, and four bulls; of which Ebor, a yearling sold for 90 guineas. The sale of this extraordinary herd realized a total amount of £4,558 1s.; and great as this sum may seem, it is not in any degree extravagant to suppose that had the identical animals been in existence in 1839, and put up for sale after Mr. Bates's unparalleled triumph as a breeder of short-horns at the show of the Royal Agricultural Society of England at Oxford, in obtaining four principal prizes with the only four animals entered by him on that occasion, the sixty-eight head of cattle would then have realized double the sum they did on the 9th inst. In support of this opinion the writer can state upon undoubted authority, that so great was the estimation in which the premium animals referred to were held, that an offer of 400 guineas each for the premium cow and heifers was refused; and that for the bull Duke of Northumberland, Mr. Bates might have had almost any sum he might have asked; but he considered the animal valuable above all price. When the circumstances of the great yearly increase and the diffusion of short-horns of the very first class in every part of the kingdom for many years past, and the crushing influence which Free-trade policy must have on the price of cattle, are considered, the proceeds of Mr. Bates's herd fully corroborates the writer's opinion of its being the most excellent ever submitted for sale by auction."

Remember that every horned animal on the farm was exhibited and sold, and there was not one kept back. The Booth sale in Sept., 1852, 44 lots averaged only £48 12s. 8d. (Dixon, page 199), and then, of course, no inferior stock or store stock were sold.

The parties who wish to deprecate the stock of Mr.

Bates, often ask if they were so superior to other herds, why did they not sell for greater prices? At the time of the sale, 8th May, 1850, the agricultural interest, in all branches, was in a very depressed state. The shock caused by the Free Trade had paralized the spirit of all classes, and corn and stock were at the lowest ebb. I quote the statement of a well-known writer on the sale:—

"At Mr. Bates's sale Lord Ducie was as undaunted as ever, and it was nothing but being in racing phrase 'a good beginning,' which secured him the 4th Duke of York so cheap. He had determined to buy him, or make him dear for some one, and he put in so promptly at 200 guineas, that, although one gentleman, at least, wished to have him at 200 more, a sort of stagnation supervened, amid which Mr. Strafford's glass ran down. If the first bid had only been 100, three at least would have gone in."—*H. H. Dixon*, page 156.

Mr. Thomas Bates, of Heddon, has sent me the following observations on this sale, for which he vouches, and is alone responsible. "In 1850 I knew nothing about the value of short-horns, or the mode of selling them. As my brothers were abroad, I took the management of the estate at Kirklevington. I was resident in London, and had other occupation. I was told that Lord Ducie and the Americans would give fabulous prices for the Duchess blood, and that one bull would go for 500 guineas at least, and that Mr. Bates had valued him at 1000 guineas. I offered to sell the whole herd to Lord Ducie, and wrote his lordship a letter with my offer to negociate. His lordship wrote declining, and at the same time a letter was written in his bed-room that I should be watched, and my plans disclosed by a person in Lord Ducie's confidence, who undertook to report all he could learn from me, and that his letters doing so were written at Lord Ducie's bedside in London, he being then confined to bed by illness. When the above person called on me at my chambers, at Lincoln's Inn, to watch me as he had undertaken, I put a copy of his letter written at Lord Ducie's bedside into his hand, he read it, and sneaked off. He trembled too much, and looked too contemptible to kick out."

"I never suspected that an English nobleman, much less one of Lord Ducie's character, could be a party to any sporting or racing tricks to obtain cattle on any terms. But I learnt, afterwards, that it was well known in the short-horn world, that Lord Ducie was determined to have the stock, and that to "bid against him was only taking money out of his pocket;" and that his lordship, who was well known and popular with breeders, personally attended the sale, that the 'undaunted' course might, most surely, have 'a good beginning.' Hence the bid of 200 guineas, and others, by his lordship in person. I travelled by night from London, and was present. I felt the pause at the sale, and the sensation made by this bid, and was only restrained from bidding by the undertaking that the sale was without reserve. The Americans, who I expected to buy the Duchesses, were not at the sale. I was afterwards informed that there was some misunderstanding in America about the period of the sale. I had not then read Mr. Vail's letters respecting the appreciation of the stock in America."

"Duchess 55th was sold to Lord Ducie, as served on the 1st of April, 1850, by 2nd Duke of Oxford, but she calved on 25th October, 1850, to the 4th Duke of York. This calf was Duchess 66th, whose portrait appears in the

Herd Book, vol. xi., page 415. So that the Earl of Ducie by this racing proceeding got one bull 300 guineas under his real auction price, and a cow in a proper breeding state, instead of a doubtful one, and the calf sold for above 700 guineas."

"When a resident Fellow in College at Cambridge, I had taken agricultural friends to visit Mr. Jonas Webb's stock at Bebraham. I found that Lord Ducie and Mr. Webb had been at Kirklevington the day before the sale, and when I arrived just before the sale, Mr. Webb introduced me to Lord Ducie, and I paid them all the attention I could. His lordship was suffering from gout at the time, and could scarcely walk. I had always looked upon Mr. Webb as a man who would not lend himself to any racing tricks. I did not then understand cattle selling, and I did not think it necessary to consult any friend on the subject. I had known Mr. Crofton and Mr. John Wood, and I once intended to have asked their advice and assistance. Mr. Webb might not know what Lord Ducie certainly knew on his sick bed, that I had put no reserve, and had taken no steps to counteract their racing schemes."

"I well recollect when the old cows were knocked down at about 20 guineas. There was a great commotion at one side of the ring. Some laughed ironically, and others raised a sort of cheer, and there was a cry of 'fat price—butcher's prices.' I afterwards ascertained that there was a strong party of breeders who had long smarted under the observations of Mr. Bates on their herds, and on the alloy hard handling, lack of quality, training, &c., who attended the sale in large numbers, and they could not restrain the triumph they expected in the herd being sold for butcher's prices, and that Mr. A. Maynard, much to their disguist, soon upset the party by his spirited biddings."

Among the English noblemen who have been distinguished as agriculturists and breeders of stock, no one ranked higher than the late Earl Ducie; but like every one who ever tried short-horn breeding, he became an enthusiast and devoted himself perhaps more than Earl Spencer to the subject. As soon as Duchess 55th was knocked down, he said I will put her to Userer and improve her shoulders. This he did, and she produced a white heifer calf; and as soon as his lordship saw it, he sent for Mr. Strafford, and when he arrived his lordship said, "Well, Strafford, Bates is right and I am wrong. I will never cross them again with anything except themselves," viz., the Duchesses and the Oxfords.

His lordship sent his agent to purchase the 3rd Duke of York, with instructions to sell him to a butcher and see him killed. The purchase was made but no butcher could be found, and he was taken to Tortworth and killed. His lordship supposed he had thus secured the only remaining bull of this blood, but meeting Mr. Tanqueray in London, he asked him what he was doing in the short-horn line, He replied " I have just come into possession of the 5th Duke of York," to which his lordship replied, " d—— that bull, I had lost sight of him." Probably his lordship would have done wisely if he had retained the 3rd Duke in his herd.

TORTWORTH COURT, GLOUCESTERSHIRE,

Sale of Duchesses and Oxfords, the property of the late Earl Ducie, on the 24th August, 1853, was the next public appearance of the Kirklevington herd :—

Lot.	By whom purchased.	Price £ s. d.
4	Duchess 55th, Mr. Tanqueray, Hendon,	52 0 0
11	Duchess 59th, Mr. J. Thorne, New York, U.S.,	367 10 0
20	Duchess 64th, Do., Do.,	630 0 0
27	Duchess 66th, Messrs Becar & Morris, New York, U.S ,	735 0 0
35	Duchess 67th, Colonel Gunter,	367 10 0
40	Duchess 68th, Mr. J. Thorne,	315 0 0
44	Duchess 69th, Mr. Tanqueray,	420 0 0
47	Duchess 70th, Colonel Gunter,	325 10 0
	Average,...£401 11 3	£3212 10 0
10	Oxford 6th, Mr. Tanqueray,	215 5 0
21	Oxford 11th, Do.,	262 10 0
37	Oxford 15th, Earl of Burlington,	210 0 0
46	Oxford 16th, Mr. Tanqueray,	189 0 0
	Average,...£219 3 9	£876 15 0
1	Duke of Glo'ster (11382), Messrs Becar & Morris, and Mr Tanqueray,	682 10 0
2	Fourth Duke of York (10167), Gen. Cadwallader, Philadelphia, and Mr G. Vail, Troy, New York, U.S.,	525 0 0
9	Fifth Duke of Oxford, Lord Feversham,	315 0 0
	Average,...£507 10 0	£1522 10 0
	Total Average,	£371 2 4

I cannot enumerate all the sales, or trace all the Kirklevington herd. I mention a few, and as the public accounts of such sales contain often a history of the animals I cannot do better than transcribe from them. Mr. Harvey Coombe's Cobham herd had been sold at large prices:—

"At Mr. Hale's sale there was not a large but a very choice company of breeders. Moss Rose, the 260 guinea calf at Cobham, made 245 guineas, after a fine contest between Mr. Robarts and Mr. E. L. Betts, the buyer. The fight for the dashing Bates Bull, Fourth Duke of Thorndale, was short, sharp, and decisive between Captain Gunter and Lord Exeter's agent, the Captain putting him up at 200 guineas, and shooting his opponent 40 guineas at a time till 400 guineas was reached, when the latter came again with his 'and ten,' and secured the American roan for Burleigh. He earned 250 guineas during his first season in England."

This was, I believe, the first time a bull had been imported from America to England. He was bred by Mr. Thorne from his Tortworth purchases, and hence the name.

The next sale called forth the above leading article in the *Times* newspaper. It astonished not only London, but the world. I have endeavoured to show how Mr. Bates fulfilled all the conditions that the writer thought essential to produce such animals—talent, education, industry, and the devotion of his time and fortune to the pursuit. The following account is from the *Illustrated London News*, a paper which seldom appears without a mention of the Bates blood :—

THE LATE LORD FEVERSHAM.

[I consider it due to his memory that the following should appear in this history :—

"Short-horns had no more devoted lover than the late Lord Feversham. He kept a large herd of them at Duncombe Park, and from a dozen to a score at Oak Hill, near Barnet, to amuse his town leisure. Whenever there was a great sale he was to be seen with his umbrella under his arm. When Parliament was up and he got home again to Yorkshire, he would often walk right across the park to the calf-house to see the new arrivals before he set foot inside his door. His father kept Devons up to 1818, but they did not answer, and he therefore laid himself in with short-horns at the Barmpton and

Mason's sales. Old Hannah, the little cow-woman, was a wonder in those days, no one could manage the refractory bulls as she could; and it was a grand treat to sit down in the lodge and draw her out, when she had become old, about her deep love for 'Young Grazier and the cows of her buxom womanhood, fine roomy yans,' as she always put it. The late Lord was a purchaser both at Kirklevington and Tortworth. He was a Bates man to the very core; and he presided at the Willis's Rooms sale of the Duchesses. He very much improved the milking quality of his tribes with First and Second Cleveland Lad, and at Tortworth he gave 300 guineas for Fifth Duke of Oxford, then a six months calf, and 120 guineas for his half-brother, Gloucester. The former won the first Royal aged-bull prize, thanks to his fine handling, at Chester; and Gloucester won a head prize at Paris. His Lordship had acted both as president of the Royal Agricultural Society and of the Smithfield Club; and, in a less conspicuous way, he may very well be ranked with the Duke of Richmond and Earl Spencer."]

"Willis's Rooms have been put to many and multifarious uses. Pictures, virtu, and china, have all made fabulous prices in them; but 'the intelligent foreigner,' if he does not happen to be a short-horn man, will be sorely puzzled when he hears that, on June 7th, Mr. Strafford will sell there a number of Grand Duchesses and Grand Dukes. Frenchmen may become puzzled about 'Sir Strafford,' and confound him with the whilom Sir Strafford Canning, or falling back upon their original idea of our wife-selling propensities, conclude that the Grand Dukes, deterred by a more civilized code from getting rid of their spouses on the Continent, have come over in the height of the fashionable season, to do it under the shadow of our English laws and customs. Happily, no foreign states will be bereft of their 'old nobility,' male or female, by the step; but the Dawpool herd will be simply dispersed. The executors reserve to themselves the right of selling the stock in one, two, or more lots. With the exception of Imperial Oxford (18084) the entire herd is descended from the Bates-bred Duchess 51st. A perfect bridal lunch greeted the congress of about 120 leading short-horn men—pens, M.P.'s clergymen, and laymen—who attended to see the great battle at Willis's Rooms, over the eighteen Grand Dukes and Duchesses. Lord Feversham was in the chair, supported by General Hood (who came, like several other members of council, direct from Hanover Square), and the Bates men made up a most imposing array; while Mr. Torr and Mr. Thos. Booth were at the head of the great rival house of 'the red, white, and roan.' As soon as lunch was over, and the noble chairman had given the usual loyal toasts, Mr. Strafford responded to his own health in the most practical way, by shutting the windows to get rid of the street roar, mounting a table, and proceeding to business, and, we may add, without repeating the new sale creed, which, by the bye, seems most searching, and nearly a foot long. The herd were at Dawpool, where they have had some crowded levées for several days past. It was their late owner's wish that they should be sold in one lot, and it was understood that the Hon. Col. Pennant, M.P., had made an offer of £6,000 for the twelve females. Aided by Mr. Drewry, agent to the Duke of Devonshire, who was said to be bidding on behalf of an eminent city merchant somewhat anxious to begin a herd, and Captain Oliver, who was in at 1800, the biddings fairly flew for the first lot; but Mr. Betts' agent was always there, and it fell to him for 1,900 gs. The second lot followed suit at 1300 gs.; and then the opposition made another rally, but at 1800 gs. Mr. Betts covered them again, and when it was seen that he was so determined, no one would bid against him for the fourth, and his 1200 guinea bid remained unchallenged. Their new owner was sent merrily along for Imperial Oxford up to 450 gs., and then Grand Duke 6th (130 gs., Mr. Bland), Grand Duke 13th (310 gs., Mr. T. Walker, of Buswell Hall, near Coventry), and Grand Duke 10th (600 gs., Duke of Devonshire). The waiters, whose ideas did not go beyond milch cows standing among ginger-beer bottles in the Mall, stared with amazement again; and the trainer of the Duchesses was so delighted at the last price that he rose and asked the auctioneer, 'how much for the depreciation in short-horn prices?' an allusion which was caught up and highly relished. Grand Duke 9th and Grand Duke 15th were confessedly rheumatic; but Captain Gunter gave 100 gs. for one, and the other, which is said to be a hopeless case, was left unsold."

SALE OF MARQUIS OF EXETER'S HERD, MARCH 14TH, 1867.

"The weather prophets have shared the usual fate of the racing ones, and have been 'floored to a man' over this fitful March weather, which has nearly stopped hunting and ploughing everywhere. However, neither snow nor cold damped the ardour of the 'short-horn parliament men,' when they met on March 14th, 1867, at the Marquis of Exeter's home farm, with Mr. Strafford as 'Mr. Speaker.' The 'Burleigh nods' from the Commissioners of Lord Penrhyn and Captain Gunter had a tremendous ten guinea significance, when that 'humble, but meritorious, occupant of our pastures' (as an M.P. recently observed to the House), Fourth Duke of Thorndale (17750) was led into the ring. This rare eight-year old bull was imported from America, in 1861, and sold to Mr. Hales, of North Frith, in Kent, who re-sold him, the following year, to the late Marquis of Exeter, for 410 guineas. He is still very fresh and active, and one of the very best bred Bates bulls in the world. His sire, Duke of Gloucester (11382) was knocked down for 650 guineas, at the Tortworth sale, and his dam, Duchess 66th, for 700 guineas. The latter is also the dam of Captain Gunter's Duchess 70th, another 310 guinea purchase on that eventful day. Fourteen years had not taken the fire out of the Captain's bids, and the bull went to Wetherby Grange at 440 guineas, or a 30 guineas on the late Marquis's purchase-money. This is another illustration of the saying that 'really good blood will always pay.' Such a figure for 'The Fourth Duke' is a rare omen for 'May morning' (new style) near Maidstone. The biddings at Burghley, on behalf of Lord Penrhyn, arose from his lordship's wish to have a worthy successor to his celebrated prize bull Duke of Geneva (19614), who died at Penrhyn Castle home farm, on the 26th February, 1867, from a tumour, which, for five or six weeks, had been perceptibly closing his gullet. He was just seven years and ten days old, and was by Second Grand Duke (12961) from Duchess 71st. His price, on his arrival from America, was 600 guineas."

THE PRESTON HALL SALE OF SHORT-HORNS, MAY 1ST, 1867.

"The breeders of short-horns, and especially the admirers of the Bates blood, enjoyed a rare treat when the magnificent herd of

Mr. Edward Ladd Betts, Aylesford, Kent (comprising sixty-two animals—fifty-two females, and ten males), was brought to the hammer by Mr. Strafford. The Preston Hall sale adds a fresh page to the history of the Bates short-horns, which will be frequently referred to with just enthusiasm as a proof that, despite the difficulties attending high breeding, and notwithstanding the extending distribution of the best blood, short-horns, *when bred with judgment*, will still maintain their price. We believe that the Kentish sale, taking into account the number sold, presents the highest average yet realised—viz., a total of £11,187 5s. or £180 8s. 9¼d. per head. This surprising result is mainly, of course, attributable to the presence of the Grand Duchesses, whose remarkable history stands out as a proof of the value of purity of blood.

"The grand heifer selected by Mr. Bates at Colling's sale with such unerring judgment has indeed proved the mother, not 'of a race of Queens,' but of Duchesses and Grand Duchesses, unequalled for aristocratic character and fine breeding. Long may they flourish, and may their present owners bring to bear upon their future development as much judgment as those who have hitherto possessed them. The Grand Duchesses result from Duchess 51st, purchased for 50 gs. at the Kirklevington sale in 1850, by Mr. Bolden, who, after a period of ten years, parted with the thirteen animals comprising the family to Mr. Hegan, of Dawpool, Cheshire, for the round sum of 5000 guineas. At his death, seventeen animals were disposed of at Willis's Rooms for 8000 gs.—eleven females and a bull going to Preston Hall, at an average of £512 or thereabouts; and it cannot be doubted that a large portion of the numerous company, who flocked round the ring, were attracted by curiosity to see if animals so costly could maintain their price. Their most ardent admirers must have been satisfied, taking into account all the circumstances of the sale. Be it remembered also that two females were lost, that Duchess 8th was a doubtful breeder, and that the most promising bull-calf, Grand Duke 16th, perfectly well on Friday night, was found dead in his stall on the Saturday morning from Hoven; a most unfortunate loss, as the 500 gs., which he most certainly must have made, would have raised the average from £442 17s. 3¼d. to over £500. Many, no doubt, were drawn to the sale by a belief, which the result has justified, that it would be the last opportunity of seeing the family intact. A glance at the subjoined tables will show how they are distributed. The ball was opened by a bid of 100 guineas for Grand Duchess 5th, a truly grand cow, nine years old, which had bred two calves since her residence in Kent, and whose offspring, Grand Duke 5th, was sold when fourteen days old, for 500 guineas. She was served by Lord Oxford 2nd (20215), and fell to Mr Drury, for the Duke of Devonshire, at 200 guineas. Grand Duchess 8th, another rich red—a year younger—which had recently calved, was evidently much admired and highly esteemed. Opening at 200 guineas she rapidly rose by fifties and hundreds to the respectable total of 550 guineas, at which last quotation the fortunate man was Lord Penrhyn, who, I believe, purchased her calf at Willis's Rooms for 500 guineas, and his affection for the dam spoke well for the progress of the youngster.

"Grand Duchess 9th, a red roan, by Grand Duke 3rd, (16182), the mother of two splendid heifers, to be described anon, had unfortunately lost her calf, and being only bulled a few days, was the victim of some uncertainty, and accordingly was run down for the comparatively small sum of 210 guineas. She was a very splendid animal, compact and lengthy, with a beautiful head.

"Grand Duchess 11th, another roan, own sister to the last, and a year younger, a six-year-old, and newly calved. The biddings for her were simply 200, 400 guineas. The buyer was Lord Spencer, whose name elicited a hearty cheer round the ring. She was quite a treat to look at, and would repay a careful examination.

"Grand Duchess 15th, a four-years-old, which had never had a calf, with a steery head and thick horn, was more than a doubtful breeder, and accordingly was a slow sale at 80 guineas.

"Grand Duchess 17th, red and white, two and a half years old, and just calved, and as far as memory serves, was the highest price ever made for a female at a public auction. She was by Imperial Oxford, out of the 10th Grand Duchess; she had all the makeness of a good one, with long characteristic hind quarters, and splended ouch. It was evident from the first bid that she would not be given away. 300, 400, 500, 600, and 700 guineas were announced in quick succession, amidst prolonged cheers, and then the glass ran; and, more slowly, 750, 800, and finally 850 guineas was debited to Capt. Oliver, whose pluck or necessities were in this case greater than Lord Spencer, the noble duke, and others. Such a scene brought to mind the disposal of the Duchesses at the far-famed Tortworth sale, when ten animals averaged £442 1s., the closeness of the two results being remarkable. Grand Duchess 18th, calved on the same day as the last, and out of Grand Duchess 9th, was of a less fashionable colour, and not so level a cow, but forward in calf. She fell to the same bidder, after a spirited contest, for 710 guineas, so that the pair realised the fabulous total of £1,638.

"Grand Duchess 19th, by Imperial Oxford, a red yearling, very large, very straight, with remarkably good loins, and altogether a thick-fleshed animal, was duly appreciated, and after another spirited fire, was dropped at 700 guineas to Mr. Dawson, of Weston Hall, Yorkshire.

"The most amusing feature of the sale was when Grand Duchess 20th, a calf under eight weeks old, was turned loose in the ring—a baby, though it must be confessed a very sweet and promising one; and whilst it frolicked in happy ignorance, was the subject of biddings of fifties and hundreds. Roars of laughter hailed the advent of each successive bid, it seeming absurd to the outsiders; but all things have an end, and so at last the glass ran down at 430 guineas, and Lord Spencer's name was again hailed with applause.

"Grand Duchess 21st, the last, a red calf five days younger, not to be compared, however, though straight and promising, was secured by Mr. McIntosh for 330 guineas, the ten females averaging £468 2s. each.

"Owing to the loss already alluded to, only three males appeared. Of these Grand Duke 4th (19874), six years old, was bought by Lord Spencer for 210 guineas. He was the joint property of Mr. Betts and Mr. McIntosh, and had been principally used by the latter.

"Grand Duke 16th, out of Grand Duchess 8th, a rich red, under two years, with immense length of body, was put up at 300 guineas, and rapidly reached 510 guineas, at which sum he was handed over to Mr. Robarts.

"Lastly, Grand Duke 17th, a rich roan, by Second Duke of Wharfdale, and a grandson of Grand Duchess 9th, with rather low shoulders, long carcase, and great hind-quarters, reached by somewhat slow degrees the respectable sum of 305 guineas, bid by Mr. Brogden. The three bulls thus made a total of £1076 15s., or £358 15s. each, the average of the thirteen, as before stated, being £442 17s. 3d. All, whether followers of Bates or Booth, must or should rejoice at such splendid results; and if it be permitted, for the departed to be cognizant of mundane affairs, then must the spirit of the founder of the Grand Duchess and the Cambridge Rose tribes, have surveyed the scene with considerable satisfaction. Scarcely less successful were the ten descendants of Cambridge Rose 6th, bought by the late Harvey Coombe, Esq., at the Kirklevington sale, the two older cows, Moss Rose and Red Rose, being by Marmaduke (14897), the remainder by 4th Duke of Thorndale (17750), and Grand Duke 4th (19874); thus securing a combination of the Cambridge Rose and Duchess blood, resulting in some very showy level cows. Thorndale Rose fell to Mr. Adams, of blood manure celebrity, which caused some merriment. Generally rich reds and roans, they had sweet breedy heads, with a deal of substance and good quality.

GRAND DUCHESS TRIBE.

		£	s.	£	s.
Grand Duchess	5th...Duke of Devonshire	210	0		
Do.	8th...Lord Penrhyn	575	10		
Do.	9th...Mr. F. Leeney, Kent	220	10		
Do.	11th...Earl Spencer	420	0		
Do.	15th...Mr. Miller, Carmarthen	84	0		
Do.	17th...Capt. Oliver, Northampton	892	10		
Do.	18th... do.	745	10		
Do.	19th ..Mr. Dawson, York	735	0		
Do.	20th ..Earl Spencer	451	10		
Do.	21st ...Mr. McIntosh	346	10		
Grand Duke	4th...Earl Spencer	220	10		
Do.	16th...Mr. Robarts, Bucks	535	10		
Do.	17th...Mr. Brogden, Cumberland	320	5		
				5757	5

CAMBRIDGE ROSE TRIBE.

		£	s.	£	s.
Moss Rose	Mr. Davies, Cheshire	241	10		
Red Rose	Mr. C. Leeney	346	10		
Thorndale Rose	Mr. Adams, Fenchurch-st.	372	15		
Red Rose 2nd	Mr. J. J. Stone, Monmouth	86	2		
Do. 3rd	Mr. C. Leeney	157	10		
Do. 4th	Mr. Brogden	252	0		
Moss Rose 2nd	Mr. Forster, Cumberland	168	0		
Thorndale Rose 2nd	Lord Braybrook	210	0		
Cambridge Duke 2nd	Mr. C. Leeney	220	10		
Royal Cambridge 2nd	Mr. Mace, Gloucestershire	94	10		
				2149	7
			Total	7906	12

SALE OF SHORT-HORNS AT HAVERING PARK.

"The sale of Mr. McIntosh's herd was quite as great a success as that at Preston Hall. It is true, there lacked the extraordinary interest attending the distribution of the Duchess tribe; but a very business-like company assembled, and expressed great satisfaction at the general excellence of the stock, and the fine healthy condition in which they were shown. Mr. McIntosh has been a close follower of Mr. Bates, to whose advice in early life, he owes, in some measure, his choice of originals, and, looking at the general quality of the herd and the very high average reached, it is evident good judgment has been exercised. The most prominent families disposed off were the Oxford and Waterloo. Third Duke of Thorndale (17749), and Fourth Grand Duke (19874), had been largely used for some years, and both had done good services. The calves by the 4th Grand Duke were a credit to him, and he might have realised a higher figure had he been sold here, in the midst of his offspring, instead of at Preston Hall. There were only four specimens of the Lady Oxford blood, from the American cow, Lady Oxford 4th. As regards the biddings for the first of these, Lady Oxford 5th, she was put up at 300 guineas, and attainted her 600 guineas, at twenty-seven bids, the last of which was a 25 guinea one. She was by 3rd Duke of Thorndale, and was the winner in the calf-class at Worcester, was a splendid roan, very level, with great substance and quality. The last bid of 25 guineas was by the Duke of Devonshire. Lady Oxford 6th, her younger sister, was put in at a reserve of 1000 guineas, Mr. McIntosh having claimed that privilege, and remained with the Grand Duchess calf, purchased at Preston Hall, to commence another herd. Baron Oxford, a two-year old, by Duke of Geneva (19614), had evidently been fixed upon by more than one, and was bought by Col. Towneley, for 500 guineas; his half-brother, by Grand Duke 4th, not a year old, was a very taking animal, and opinion varied as to which was best; he went to Dumbleton, at the same figure. The three animals sold reached the enormous average of £560. It should be mentioned, in reference to the Preston Hall sale, that Lord Oxford 2nd, a very grand bull of this blood, imported from America, was kept back on account of lameness, or the bull averages would have been materially increased. The Waterloo tribe was well represented, and found ready admirers, the eight females and one bull, averaging £122 12s. 4d. Never, perhaps, were such important sales held at a more unfavourable time, during the time of the cattle plague, yet, the result of pluck and judgment has been a higher total than has been hitherto reached."

LADY OXFORD TRIBE.

		£	s.	£	s.
Lady Oxford 5th	...Duke of Devonshire	630	0		
Baron Oxford	........Col. Towneley	525	0		
Baron Oxford 2nd	...Mr. E. Holland, M.P.	525	0		
				1680	0

WATERLOO TRIBE.

		£	s.	£	s.
Waterloo 24th	........Mr. Z. Walker	110	5		
Wellingtonia	........Mr. Davies	126	0		
Do. 2nd	...Duke of Devonshire	168	0		
Do. 3rd	...Mr. Maskell, Yorkshire	58	16		
Do. 4th	...Mr. Startoris	115	10		
Do. 5th	...Mr. Whitworth, Derby	147	0		
Do. 6th	do.	131	5		
Do. 7th	...Lord Penrhyn	126	0		
Wellington 4th	......Mr. Kettley	120	15		
				1103	11
			Total	£2783	11

Our amusing friend *Punch* had his attention called to the short-horns by his ever watchful would-be Bulldog *Toby*, and on the 2nd November, 1867, came out with the following report of the sale at Windsor:—

THE GOLDEN SHORT-HORNS.

Mr. Strafford raised his time-glass, and Thornton held the pen,
When to a Windsor coffee-room flocked scores of short-horn men.

They crowded round the table, they fairly blocked the door;—
He stood Champagne did Sheldon, of Geneva, Illinois.

They talked of Oxford heifers, Duchess bulls, and how the States
Had come into the market with another 'Bit of Bates.'

Their expression is so solemn, and so earnest is their tone,
That nought would seem worth living for but 'red and white and roan.'

All ready for the contest, I view a dauntless three—
The McIntosh from Essex, a canny chiel is he.

There's Leeney from the hop yards; 'twill be strange if he knocks under,
When once the chords are wakened of that Kentish 'Son of Thunder.'

The Talleyrand of 'trainers' is their 'cute but modest foe,
Him whom the gods call 'Culshaw,' and men on earth call 'Joe.'

And sure, it well might puzzle 'the Gentleman in Black,'
When the three nod on 'by fifties,' to know which you shall back.

And sure, the laws of nature must have burst each ancient bound,
When a yearling heifer fetches more than seven hundred pound!

Bulls bring their weight in bullion, and I guess we'll hear of more,
Arriving from the pastures of Geneva, Illinois.

SALE OF DUCHESSES AND OXFORDS, BELONGING TO MR. SHELDON, OCTOBER 19TH, 1867.

"The three Duchesses and six Oxford short-horns, two bulls and seven females, which arrived from America, and did quarantine in the 'Nestor,' were on view at the Shaw Farm; but the biddings for them were conducted at the Castle Hotel. The coffee-room was full of 'familiar short-horn faces;' and George Fordham the jockey, must have wondered what was up when pedigree allusions were made to 'Duchess of Fordham,' own sister to Captain Gunter's 'Duchess 70th.' Mr. Leeney, of Kent, was the chief bidder, and did it with an emphasis and courage which added great zest to the proceedings. Mr. McIntosh was also highly pugnacious, in the best sense of the word; he led off with the 3rd Duke of Geneva (550) and his remorseless 'and ten,' for nearly all the other lots fell like sweet music on every ear but Mr. Leeney's and Mr. Culshaw's. Mr. Sheldon's agent had a 600 guinea reserve bid on 7th Duchess of Geneva, but he did not exercise it, as a spirited rally between Kent and Essex made her Mr. Leney's at 700 guineas. 4th Maid of Oxford (Mr. Leeney, 300 guineas) 5th Maid of Oxford (Mr. Downing, 200 guineas), and Countess of Oxford (Colonel Kingscote, 250 guineas) followed; and then Mr. Culshaw, took 6th Maid of Oxford (400 guineas), and 8th Lady of Oxford (450 guineas) for the new Towneley herd. 7th Maid of Oxford (260 guineas) completed the Leeney trio; and then Mr. E. Thorne, of America, gave 185 guineas for 12th Duke of Thorndale (a bull bred by his brother, and looking very much out of sorts after his passage), 'rather than see him go under his value.' It was said that 600 guineas had been refused for the 3rd Duke of Geneva in America. The nine averaged £384 8s. 4d., and set the seal on a very unique and successful day. People differ in opinion as to whether the American lots would have made most under the greenwood or round the mahogany tree. 'The sea-sick bull' should almost be excluded from the average; and as, owing to their being so young (two of them were under sixteen months), only one of the seven heifers could be put down as 'safe in calf,' and four of the other had to date their 3rd Duke of Geneva hopes from September 23rd, or October 2nd, the average of £408 3s. 9d. for eight, was a large one. Much of it was owing to Mr. Leeney's Kentish spirit, as he bad well for every lot. He is a good bidder for a sale, but not for himself, as his dashing, defiant style rouses men with long purses and less decision. This style of bidding does better in a room than in a ring, where antagonists are not within ear-shot and do not get their mettle put up so effectually; and therefore Mr. Sheldon, perhaps lost nothing by the adjournment from the Shaw Farm to the Castle Inn. One thing was certain, the ring was too large and the room too small, but still large enough for those few who meant business. The year 1867 was an *annus mirabilis* in short-horn history. Mr. Betts had the largest average on record; £180 19s. for sixty-three, and £116 12s. 6d. for fifty-seven at Mr. McIntosh's. The price of Grand Duchess 17th (850 guineas) was the largest ever given for a female at a public sale. Grand Duchess 18th (710 guineas), Grand Duchess 19th (700 guineas) 7th Duchess of Geneva (700 guineas), Lady Oxford 5th (600 guineas), and Grand Duchess 8th (550 guineas), which made an average of £719 5s. for half-a-dozen. The highest bull price has been 550 guineas for 3rd Duke of Geneva, and four bulls averaged £540 15s.

I fear I have omitted to mention Mr. Atherton in connection with the Kirklevington herd. His farm and stock are very fully described by Mr. Dixon (page 376), and I venture to extract from his work. Mr. Atherton's farm, Chapel House, lies five miles from Liverpool. It has been originally a rabbit warren, but "Time works wonders," and in 1857 three societies awarded Mr. Atherton a prize for the best cultivated farm, and almost innumerable prizes and cups for crops of various kinds. His great short-horn sale was in July, 1862. Three Grand Dukes made £750 15s., the 7th by the 3rd from Grand Duchess the 4th, going at 320 guineas to Captain Oliver, and the 3rd at 195 guineas to his neighbour Mr. Robarts. Six Cherry Duchesses made £696 3s., and of these Cherry Duchess the 7th (205 guineas) went to Lord Penrhyn.*

* Mr. Atherton has both bred largely and had several short-horns of much value through his hands. The brothers Robert and Thomas Bell, who had left Kirklevington after the great sale, came to reside in his neighbourhood, and gave him the first start with pure blood. A lot of heifers, with Marquis of Speke (13307), and then Cherry Duke 2nd (which he got from Mr. Bolden) to serve them, brought the herd to 50 strong, and these were sold in March '58 for

In writing of the Duke of Devonshire's herd at Holker, Mr. Dixon says:—

"The Cozies and the Nonsuchs were given up and the Grand Duchess (1), Oxford (8), Wildeyes (7), Blanche (8), Barrington (2), Gwynne (1), Oxford Rose (3), Cleopatra (2), and Waterloo (1), were al represented. Eighteenth Duke of Oxford was the last hope of the calf-house, and Lady Oxford the 5th (the 600 guinea Royal Worcester calf) was on the eve of calving her fourth Baron Oxford to' that grand old Duchess bull, 7th Duke of York.

"Two of her calves were sold for 500gs. each at Mr. McIntosh's sale, and the other for 250gs. to Lord Kenlis, at Killhow. Countess of Barrington 4th somewhat reminds us of Duchess 77th, and has a son by Tenth Grand Duke at her side. The light roan is Blanche 3rd, grand-daughter of old Sylph; and a broken horn marks Seventh Grand Duchess of Oxford, who also rejoices in a beautiful-haired daughter, the 12th of that line. Lady Oxford 5th is the queen of

nearly £33 a piece. Cherry Duke 2nd headed the bulls, and was bought by Mr. George Shepherd's son, a mere boy, for 205 guineas; he did good service at Shethin, and went thence to Rossie Priory, Inchture. Mr. Atherton's second start was with Gwynnes from Mr. Caddy, and Wildeyes from Messrs. Barthropp and Crisp; these were augmented by the Springfield Duchess, Cherry and Finella purchases. Czarovitz (17654) was bought from Knowlmere Manor. Moss Rose (which he successfully exhibited) and another heifer or two came from Wetherby, and with them the Duke of Wetherby (17753), the first-born of Duchess 77th, on hire. After a short season with him a second sale took place in '62, the 51 averaging £67 5s. 8d. Since then Mr. Atherton has never lacked a good animal. In '64 he bought Mr. Mark Stewart's two heifers of the Cherry tribe, and one of them, Southwick Cherry Flower, illustrates the 18th volume of the Herd Book. He also bought some of the Kirklevingtons, which he has recently sold at a large profit to Mr. Pavin Davies. The American bull Lord Oxford 2nd (20215) was purchased by him soon after landing in '62; after nearly four years' use he was exchanged for Imperial Oxford (18084), who died in a short time, and was replaced by Thirteenth Duke of Oxford (21604) (bred at Holker), from Killhow. The latter was sold with some heifers to Mr. Edgar Musgrove.

the field, fit to found a world of short-horns for substance and true character. Oxford Rose 2nd, by Grand Duke 4th from Rose of Raby makes a nice pair with Oxford Rose, by Baron Oxford. Old white Dustie has no heifer to perpetuate her line, and Morning Star is the last dying bequest of Lord Oxford. Fifth Grand Duchess of Oxford is a wonderful milk and butter cow. From her we pass through the park to the home of Tenth Grand Duke, who is mourning the loss of Mr. Davies' Moss Rose (who bore him Royal Chester), and we bid him be of good cheer, as he puts forth his beautiful head to greet us, and walks most vigorously the whole length of his paddock into his shed for further recognition at Mr. Drewry's hands. Third Grand Duchess of Oxford was up feeding, and Mr. Fawcett's Eliza 10th and Lady Butterfly's Duchess were in quarantine in a paddock."

I here notice Mr. Bolden and his herd at Springfield, in Lancashire. Mr. Dixon says:—

"Grand Duke 3rd by 2nd Duke of Bolton (by Grand Duke from Florence, a daughter of Mr. Richard Booth's Fame) was in residence at Springfield Hall, near Lancaster, when we first went there in '59, and so was Prince Imperial, that son of 2nd Grand Duke and Bridecake a daughter of Bridget, to whom the Grand Duchesses also owe their Booth cross. He was thick through the breast, and with well laid shoulders, and though not with quite the grandeur of some of our best bulls, a touch must be dead or saucy that did not own him mellow. We found the footsteps of Fame in the Fenella family, which sprang from her daughter Fay crossed with Grand Duke. Mr. Bolden's brother purchased Mussulman, a son of Old Cherry, to take to Australia, for 150 guineas. Mr. Bolden liked the sort, and had Cherry Duchess by Grand Duke from a Cherry cow which he purchased at Mr. Lax's sale. Her son 2nd Cherry Duke was sold to Mr. Shepherd of Shethin. Mr. Bolden also had the Waterloo tribe, on which Mr. Bates set very great store.

"Mr. Bolden inherited his taste for short-horns from his father, who, like Mr. John Colling of White-house, and Mr. Lax of Ravensworth, caught his inspiration from the brothers Colling. He died in 1855, at Hyning, near Lancaster. No man was fuller of short-horn lore intermixed with the quaint sayings and doings of the old Durham and Yorkshire worthies. He kept a herd for many years, always sticking to the old-fashioned, roomy, heavy-fleshed cows; and hired Leonidas, Leander, and Royal Buck, and other bulls, from the Booths, in days when a man who gave only sixty guineas for a season was considered quite an intrepid character, and when Warlaby females could be had for money.

"Four of these then 'Veiled Prophetesses,' Fame, Rachel, Bridget, and Vivacity, were purchased by Mr. Bolden, soon after he commenced breeding, in 1849; and along with cows of the Duchess, Cambridge Rose, and Waterloo tribes, from Kirklevington; the Cherry tribe from Colonel Cradock, and the descendants of No. 25 at the Chilton sale, they gradually formed the herd.*

---

* Duchess 64th, the dam of Second Grand Duke, Mr. Bates did not live to see, and she was the youngest of eight which stood up before Mr. Strafford. Her dam, Duchess 55th, had been a very Barbelle in the herd world, as three of her produce were sold for 2,300 guineas, and she was both the dam and the grandam of a thousand-guinea bull. Mr. Bolden bought the first of the Kirklevington eight, to wit Duchess 51st, dam of the Fourth Duke of York, for whom Lord Ducie gave 200 gs. at the same sale, and sold, after three years' use, to the Americans for 500 gs. The salt water was fatal to him, as he broke his neck in a storm; but the change from the banks of the stately Tees to 'the gently curving lines of creamy spray,' that wash the Red Bank Farm, redeemed his dam from the curse of barrenness, which had sunk her to 60 gs.

She bred three heifer-calves, the first of which, by Leonidas, died in the birth, and the others were ushered into the yard at Springfield for us, in the shape of two roan cows, Grand Duchess and Grand Duchess 2nd, by Grand Duke. A noble pair they were. The eldest was a beautiful specimen of a 'toucher, silky hair on a nice elastic hide, with that peculiarly dainty cellular tissue between the hide and flesh. The head, too, had all the most favourable characteristics of the tribe, slightly dished in the forehead, with a prominent nostril, and a great general sweetness of expression. They were also well down in the twist, and great milkers, combined with heavy flesh. Grand Duchess 2nd bore a strong family likeness to her sister, but she had more substance and gaity of carriage; and she held up her head, as if right conscious of her lineage.

Three of the heifers were red with a few patches of white, and it was curious to notice in their marks the exact resemblance to that original Duchess colour. Coates's Herd Book has preserved to us her picture, as she feeds on the Tyneside, with Halton Castle in the distance. The white patch on the flanks and crop, the star on the forehead, and the gay little beauty-spot just above the muzzle, are all there. There was no break in the "red and white" succession till Duchess 19th was crossed with Belvedere of the "White bull," or the Princess family, and two roan heifers were the produce. A double cross of Belvedere brought the colour to white for the first time in Duchess 50th from Duchess 38th, by the Duke of Northumberland, from the first roan, Duchess 33rd.

Cambridge Rose 5th, by 2nd Cleveland Lad, was five years old at the Kirklevington sale, when Mr. Bolden, senior, bought her, and with the exception of Cambridge Rose 6th, who was kept as a memento at Cobham, and Cambridge Rose 7th, which was purchased by Mr. Downes (and died in '67), and from him by Mr. Bolden for 70 guineas, the next autumn, there were then no more descendants in the land of the celebrated Hustler's Red Rose. Cobham proved the value of this blood by the biddings for the gay old cow, and her Marmaduke calf, Moss Rose. The 1st and 2nd Dukes of Cambridge alone represent Cambridge Rose 7th, and as she persisted in breeding nothing but bulls. the tribe was lost to Springfield at her death.

When Mr. Bolden had got home old Duchess 51st, *and compared her* with some other very good short-horns on his farm, he became so convinced of the goodness of the Bates blood that he determined to make his stand on it. His first move was to purchase Grand Duke for 205 gs., the same price that Mr. Hay gave for him. At the time he bought him he and his father had several cows almost useless, after having been served repeatedly by idle bulls; but with him and successive Duchess bulls, the fertility (which Mr. Bates attributed, in the case of the Duchesses, to the cross with Belvedere) gradually returned. The same was observable in other herds where Duchess bulls were introduced, and Earl Ducie did not conceal his opinion that his was saved by the use of them. Grand Duke was four years old when he came, and he departed for America two years after; and whether, in addition to the Dukes of Cambridge, we look at May Duke and Grand Turk (the sire of Great Mogul), from Booth cows; and two Cherry Dukes from the Cherry tribe, all of which have been sold and resold at high figures, Mr. Bolden stands as a bull breeder second to none.

Grand Duke 2nd, by 4th Duke of York, from Duchess 64th, who was calved at Mr. Bolden's, had rather more white on him than Grand Duke, and was only two years old when he followed him, in November, 1855, to the New World. He had not quite the bold look of Grand Duke, and although it would seem to be the perfection of a short-horn to read good nature in his face, the Americans always thought that he looked too placid. Unlike the gentleman who described himself as having been absolutely unable to close his eyes from emotion, the live-long night after his unexpected " Vision of Fair Women," in the shape of Queen Mab, Nectarine Blossom, and Queen of the May, a recent visitor to Thorndale does not seem to have been the least stirred up by treading such classic soil, or much struck with anything beyond Grand Turk weighing 2,800lbs. He tells us, however, how he found him in company with 2nd Grand Duke and Neptune of the Booth blood; and how he calculates that Duchess 64th and 66th, Oxford 5th, 6th and 13th, and Bloom, Frederica, Lalla Rookh, Buttercup 2nd, Miss Butterfly, and Pearlette would be alongside them. Such an American Congress would be worth all the sea-sickness and all the expense to see. Duchess 64th (600 guineas), which was generally considered the best of the eight Duchesses that were sold at Tortworth, died after some years in America, along with Duchess 59th (350 guineas); and Duchess 66th (700 guineas), that " brand plucked from the fire" (as Earl Ducie termed her, when the news was carried to his dressing room one morning that a calf had at last been found in Duchess 55th) was among the fifty-head which Mr. Thorne purchased after poor Mr. Becar's death for £7,000.

In 1854 Mr. Bolden sold seven bulls at an average of £59 8s., one of them, Second Duke of Cambridge, for 100 gs. When Mr. Bolden, sen., died, in 1855, his herd, with some of his son's bulls, were sold at Hyning, and the 28 head (including 11 bulls) realised an average of £61 16s. 9d. Mr. Torr bought Gertrude (100 gs.), and Lady Hopetoun (220 gs.), both of them Booth cows. In 1857 Mr. Bolden sold 14, at an average of £65 3s. 4d., at Mr. Strafford's farm, at Dudding Hill. This was followed, in 1860, by a sale at Springfield, where 29 head averaged £87 17s. 6d. Of these a score were Waterloos, and they averaged £92 13s. 3d. Sir Curtis Lampson gave 165 gs. for Waterloo 20th, and Mr. E. Bowley 130 gs. for the Waterloo bull Charger. In 1862 the herd was sold to Mr. Atherton, who who soon after parted with the Grand Duchesses (nine cows and four bulls,) to Mr. Hegan, of Dawpool, by private contract, and sold off the Cherries, the Fenellas (from Booth's Fame), and the Grand

Duke bulls. Mr. Hegan paid £5,000 for his lot, and three cows were barren. He died in 1865, and his herd was brought to the hammer at Willis's Rooms."

PART XVII.

Portraits—Mr. Bates—Duchess, 1804—Ketton—Belvedere—Pedigrees—Elvira—Duchess 34th—Duke of Northumberland; sums offered for him—Wetherby Herd and Duchesses—Duchess 77th—94th and Portrait—Pedigree—My own experience and taste—Scarborough Exhibition—Mr. Cochrane's purchases for Canada—Lord Dunmore—8th Duke of York—Pedigree—Purchases by Myself and Friends—Loss of Cattle—My own Herd—Intended Sale—Conclusion.

I may here say a few words on the portraits introduced into this book. That of Mr. Bates was taken at the request of an old and intimate friend, by Sir James Ross, R.A., when Mr. Bates was about 55 years of age, and the original has been kindly lent me by its present owner, Neville Rolfe, Esq. In many respects it is a very faithful likeness.

The Duchess of 1804, or, as she is often called, the Daisy Cow, has been fully described by Mr. Bates. She does not appear in the Herd Book. I need hardly repeat she was by the Daisy Bull (186), d. by Favourite, 2nd d. by Hubback, 3rd d. (the Stanwick cow) by James Brown's Red Bull. Daisy Bull was by Favourite, d. by Punch, 2nd d. by Hubback.

Ketton the 1st I need not more fully describe. No judge of short-horns who saw him ever forgot him. He was by Favourite, d. Duchess of 1804.

I have not inserted the portrait of Belvedere, I however here state his pedigree.

Belvedere, whose portrait appears in the 3rd vol. of the Herd Book, 1836, page 21 (1706), yellow roan, calved 6th April, 1826, bred by Mr. Stephenson, the property of Mr. Bates, Kirklevington, near Yarm, by Waterloo (2816), dam (Angelina the 2nd) by Young Wynyard (2850), 2nd d. Angelina by Phenomenon (491), 3rd d. (Anne Boleyn) by Favourite (252), 4th d. (Princess) by Favourite, 5th d. (bred Mr. R. Colling, and own sister to his white bull) by Favourite, 6th d. by Hubback, 7th d. by Snowdon's bull (612), the sire of Hubback, 8th d. by Mr. Masterman's bull (422), 9th d. by Mr. Harrison's bull of Barmpton, bred by Mr. Wastell, of Burdon (669), (292) 9th d. was bought of Mr. Pickering (of Foxgall), near Sedgefield.

ELVIRA I notice as representing the Princess tribe. I append her portrait. The grandeur and style of this race no doubt was imported to the Duchess tribe by the cross with Belvedere.

PRINCESS (Herd Book, vol. I. 415) calved 1800, bred by R. Colling, by Favourite, dam by Favourite.

	Produce.	Sire.
1803	Anne Boleyn	Favourite.
1805	Elvira	Phenomenon.
1806	Wynyard	Ditto.
1809	Nell Gwynne	Ditto.
1810	Peg Woffington	Wynyard.
1813	Pilot	Ditto.

ELVIRA, (Herd Book, p. 292.)

1812	Helen	Wynyard.
1819	Constitution	Favourite (256)
1820	Elvirina	Ditto.

Note by Geo. Coates—"Good, sold to Mr. Janson, 96 guineas."

DUCHESS 34th, by Belvedere

1st dam	...	...	...	Duchess 19th.
2nd dam	...	...	...	„ 20th.
3rd dam	...	...	...	„ 8th.
4th dam	...	...	...	„ 2nd.
5th dam	...	...	...	„ 1st.

6th dam, by Favourite.
7th dam Duchess, 1804.

Her only exhibition for a prize was in 1842, when she beat Necklace. She was at York in 1848, but not exhibited for a prize.

The DUKE OF NORTHUMBERLAND, so named by Mr. Bates, "to perpetuate the commemoration that is due to the judgment and attention of the ancestors of the present Duke of Northumberland, that this country and the world are indebted for a tribe of cattle, which Mr. C. Colling repeatedly assured me were the best he ever had or saw."

"Much has been said about the sums refused by Mr. Bates for the Duke of Northumberland. He often refused 1000 guineas. One morning a gentleman came to Kirklevington and had breakfast. He then went with Mr. Bates and myself to view the herd, and just as it was concluded I was called away. When I returned, not long after, the gentleman had left. The cowman and the housekeeper then came and told me what had passed with this gentleman and Mr. Bates as they returned to the house. They were present. The gentleman said, I have come to buy the Duke of Northumberland, and am prepared to give a good price. Will you take 1500 guineas? Mr. Bates smiled, and said no; the gentleman then said 2000 guineas, still no was said. 2500, 3000, and 3500 were offered and refused, and Mr. Bates smiled more and more. The gentleman then said, I will give 4000 guineas. Mr. Bates said no, and smiled still more. The gentleman then said, will money buy him, and Mr. Bates said no. We joined in expressing our regret, and perhaps our opinion of the folly of not taking such a price, as Mr. Bates had bulls of the same blood that he could use, and in fact he did not then require to use the Duke of Northumberland."

The Duke of Northumberland gained the first prize of £30, at Oxford, July 17th, 1839, as the best short-horn bull; beating, among other celebrated bulls, the bull

Roderick Random, which had gained nine first prizes, and was never before beaten. The Duke of Northumberland was also successful on every occasion that he was exhibited: as a two-year old, he gained the highest prize at the meeting of the Yorkshire Agricultural Scocity's first show, at York, August 29th, 1838, and in September of the same year, the highest prize was awarded to him as the best bull, of any age, at the Stockton and Cleveland Agricultural Society's show at Stockton, and in October of the same year, at Darlington, the highest prize was likewise awarded to him as the best bull of any age. This bull's own sister (Duchess 43rd), gained the prize of £10, as the best short-horn year old heifer, at the Oxford meeting, as well as prizes at Stockton and Darlington the same year. The live weight of the Duke of Northumberland, July 1st (two days before starting for Oxford), was 180 stones (14 lbs. to the stone); on arriving home, July 29th, 1839, his weight was 152 stones, so that he lost 28 stones in travelling 26 days, near one-sixth of his whole weight, thus showing how much animals lose by travelling. At ten months and at two years old, he considerably exceeded the celebrated bull Comet in weight.*

Mr. Dixon in writing of the Wetherby herd (page 282) says :—Mr. Cochrane's pair have a levee in their barn

---

* E. Bates informs me that the regular loss of weight in fat cattle, from Germany to England, is found to be 14 lbs. a day, from the time they leave their stalls or pastures until they are slaughtered. This is when the journey is made by rail and steamer. This I think is a full answer to the attacks on the Duke by his detractors as to his not standing his journey, which was made on foot, 16 miles to the ship, and about 60 miles from the ship to Oxford. Could Hecatomb have made such a journey?

all day, and devotees go wandering off through the hot haze into the park to gaze on Duchess 86th, 87th, 88th, and 91st, as well as Mildeyes and her daughter Brighteyes, and a very fine Waterloo heifer:—

When we first saw the herd in '59, not long after its removal from Earl's Court, we began with the earliest purchase Duchess 67th, and her daughter 72nd, the first calf that Captain Gunter ever bred. Her next daughter the white 75th was third in the array, and the handsomest of the three, and then came "the twins" 78th and 79th which ran such a splendid career in the show yard. We see the little roan and white through the mist of years once more struggling with the herd boys, and thought the roan rather nicer in her coat, but the white neater, and in after years the bench hardly knew which to take. Having thus exhausted the fruits of the first Tortworth bid, Duchess 70th bore her witness to the second with her calves 73rd and 77th, and we look back to our comment that " the former had more substance and the latter more elegance of the twain," and that she was the best, but no one dare predict such a future for her. She rose the Royal ranks step-by-step, third as a yearling at Warwick, second at Canterbury, and first at Leeds. Duchess 69th had only calved that morning, and though we could not rouse her after the labours of the day we could judge of her fine scale and enjoy the gentle grandeur of the head, which had been specially modelled for Mr. Brandreth Gibb's testimonial. Sixth Duke of Oxford was waiting outside to receive us; he was a perfect Esau at his birth, and there could be no doubt whence his stock derived their rich hair.

["A period of nine years must be supposed to elapse," as the playbills have it.]

The old cows were in the bottom of the Park, and took a good deal of finding in the heat. There was the roan Duchess 86th, with the old-fashioned wide-spreading horn; the 87th, of a lighter roan and with a rare loin; the white 88th, which had been amiss; and 91st, one of the same colour and rare substance. The twins and the 77th had died or been slaughtered, and 96th and 94th were in the home field, and Taylor tells us how once they thought 94th the best, and that the former is the only Duchess which lacks the Usurer cross. The numbers 100th, 99th, 98th, and 97th once roamed together in the home pasture unbroken, but Mr. Cochrane had taken his choice and borne off the last to Canada at 1000 gs. She is from 92nd, a daughter of 84th, "which broke down on us as a calf for Leeds." Her once constant companion 98th from 88th was a white with roan ears. Writers who have to encounter their nightmare numbers may well be among those

"Who dread to speak of '98,
Who tremble at the name."

The wished for 100th was reached at last in the shape of a red roan, but a two-days-old roan, half-sister to "the American lady," was the latest arrival, and Duchess 103rd had been the Captain's private herd book entry. Fourth Duke of Thorndale was the monarch of the yard, and Grand Duchess 8th, from Penrhyn Castle was there to share his smiles. Mildeyes 3rd (by 4th Duke of Thorndale from Mildeyes) and a heifer by 5th Duke of Wharfedale from "the Waterloo heifer," have since then arrived; and Duchess 84th has lost the red Duchess 104th. It was jumping about its box when two months old, and burst a blood-vessel in the heart. Duchess 94th has had twins—a bull and a red heifer, the latter taking rank as Duchess 105th. Third Duke of Wharfedale (sire of Mr. Cochrane's heifer) from Duchess 86th, now reigns at Wetherby (after two seasons at Penrhyn), *vice* Fourth Duke of Thorndale, who was found dead in his box last spring; and 2nd Duke of Wetherby, from Duchess 77th, and 2nd Duke of Claro, from Duchess 79th, are both let. The 3rd Duke of Wetherby, by 4th Duke of Thorndale, from Duchess 32nd, is coming on for home use. The 2nd Duke of Collingham, Duke of Tregunter (a name taken from an old family estate in Wales) 3rd Duke of Claro, 5th Duke of Wharfedale, and 2nd Duke of Tregunter have all been sold to English purchasers for 500 gs. each.

I cannot enumerate the history of each Duchess, but I think Duchess 77th deserves notice. To Colonel Gunter, her owner, the British breeders are indebted for preserving the tribe in England, and also for keeping the blood pure. If it had not been for his spirited biddings at Lord Ducie's sale the whole tribe would have gone to America and fulfilled the prediction of Mr. Bates. Col. Gunter, in his letters to me, says:—

"Duchess 77, roan, calved Nov. 24, 1857, by Duke of Oxford, d. Duchess 70, by Duke of Glo'ster, g.d. Duchess 66, by 4th Duke of York, &c., &c. Duchess 77 won first prize as the best cow at the Royal Agricultural Society's meeting at Leeds in 1861. While I showed her she won 18 other prizes and 7 challenge cups in the counties of York, Durham, Lancaster, &c. I did not show her long as I wanted her to breed, and having produced a dead calf I never showed after. Duchess 100 was never shown."

I desired to insert for my readers' benefit a portrait of the present Duchess family, and I had fixed my attention on Duchess 100 as representing an epoch in the tribe. Col. Gunter most kindly responded to my request for a portrait. He writes me:—

"I have no pictures of the cattle that will suit you, only rough photos from nature of cattle in poor store condition. I consider some of the Duchess I have *the best I ever bred*, tho' in store condition for breeding purposes. You can have a sketch of any you like; 92, 94, 96, and 100 are probably the most likely for your purpose— especially the former, as I refused 2,000 guineas for her, and sold two heifers from her for 1,000 guineas each. I am glad the 8th Duke of York is turning out so well for you. He was a bargain, and worth nigh double what I sold him for as prices now go. The mother of your bull was a grand cow really."

My prejudice or fancy, however, for the original colour has overcome all other considerations, and I insert Duchess 94th and her calf as a proof of the transmission, not only of the blood, but of the Duchess colour.

DUCHESS 94th (Herd Book, vol. 18, page 462), rich roan, calved July 5th, 1865. Bred by, and the property of, Col. Gunter, Wetherby Grange, by 2nd Duke of Wharfdale (19649), dam Duchess 84th, by Archduke (14099) 2nd d. Duchess 72nd, by 4th Duke of Oxford (11387), 3rd d. Duchess 67th, by Userer (9763), &c.

2nd DUKE OF WHARFDALE, rich roan, calved June 20th, 1862, by 7th Duke of York (17754), dam Duchess 73rd, by 6th Duke of Oxford (12765), 2nd d. Duchess 70th, &c.

ARCHDUKE roan, calved December 23rd, 1855, by Duke of Cambridge (12742), dam Duchess 69th, by 4th Duke of York, &c.

To keep improved short-horn cattle, and test by experience the results of crosses and experiments in breeding is of course a very expensive and uncertain course for a tenant farmer.

With swine the best kinds may be obtained, and the breeds improved at no great expense, and in a short period, and in like manner with sheep. There is no loss by the milk or dairy qualities of the stock; and if they do not answer the expectations of the breeder, they can soon be made fat, and disposed of. With cattle the process is much longer, and six or seven years very often must be taken to test the merits of breeding stock and their descendants; and if the loss of the milk is to be borne, I should say that improved short-horns must prove unprofitable stock, except perhaps in America, where herds roam in thousands, and none are ever milked.

The *New York Tribune*, discoursing on farming in the West, mentions that a Mr. L. Sullivan, has, in Livingston County, Illinois, a farm eight miles square, containing 40,960 acres—64 sections, Government survey. Speaking of the immense scale on which cattle-raising is carried on in Texas, it is stated that among the large cattle-raisers are John Hittson, who has 50,000 head of cattle; William Hittson, who has 8,000; George Beavers, 6,000; Charles Rivers, 10,000; James Brown, 15,000; C. J. Johnson, 8,000; Robert Sloan, 12,000; Coggins and Parks, 20,000; Martin Childers, 10,000; and John Chisholm, 30,000. The entire number of cattle owned in Texas is nearly 4,000,000, while New York State, with her 4,000,000 of population, eight times greater than that of Texas, has less than 750,000 head of cattle.

I hope I have shown how Mr. Bates attended to this much-esteemed quality of improved stock, that not only the quantity but the richness of the milk has been duly considered.

I was born at Halton in 1805, and I was constantly with Mr Bates until his death, and my own relations and connexions were all engaged in farming, and generally with large proportion of grass land and live stock. If my taste had not been for live stock I could hardly have escaped imbibing a love for it and the enthusiasm of Mr. Bates on the subject. It was almost impossible to be with him without catching the taste.* I, however, had the taste naturally. My earliest recollections are of the best animals. I can now point out where I saw them stand and graze, and their figures remain impressed on my mind. I hope I profited by his precepts and example. To the last, he never tired of pointing out to me the points

---

* Lord Blayney relates that when travelling in Spain he received great attention and hospitality from an English gentlemen engaged in constructing the railways, and was shown a very fine large herd of cattle, in which his host took great pride. His Lordship asked how he acquired a taste for cattle. He replied—I was once cow-boy to Mr. Bates, the great short-horn breeder of cattle in England.

in cattle which were good and inferior, and the requisite touch in the various parts of the animals, and also, the great essential of milking properties. I certainly contracted from him the opinion that there was really no rivalry between him and Mr. John Booth in blood, and I trust I may be excused if I have been led away to make any remarks on the Booth stock which might imply such a rivalry. I trust I have set out the facts on which Mr. Bates justified his assertions, that he obtained the best blood the Messrs. Colling ever had. At any rate, he had all the blood from the best animals to which the Booth men refer back, and that without the alloy; and I confidently submit to the judgment of short-horn breeders that Mr. Bates was correct in what he said—he had improved upon the old celebrated animals. We must recollect that Mr. Bates criticised, as mentioned by Mr. Fawcett, the herd of Robert Colling, and at the sales of C. Colling and R. Colling it was no chance or caprice that confined his purchase to one heifer; the only female Duchess in existence. Phœnix the dam of Comet, was no doubt the finest cow the Collings possessed, and the modern Duchesses and Grand Duchesses equal or exceed her in all the essentials of dimensions.*

If the blood of the bulls in the time of the Collings produced such wonderful effects in the herds they went to,

---

* For the curious, a measurement is subjoined of the cow Phœnix. Lady Fragrant, the first prize cow at Leicester, 1868, recently measured in a reduced state, is, in nearly every point, except space and height, a larger animal, whilst the 850 gs. Grand Duchess 17th, in breeding condition, is as nearly the same size as possible. Phœnix, height, 56 in.; length quarter, 21 in.; girth at chine, 85 in.; width of hooks, 26¾; length of back, 61½; girth at neck, 38½; width of loin, 19¼; length of space, 15¼; girth at shank, 7½ in.

we may with confidence trust to the same when that blood has been preserved in its purity through so many subsequent generations, and the Americans may use this pure blood with no fear of the dreaded black nose—a stain in pure short-horns as foul as an African characteristic would be in the white human race.

It may seem strange that Mr. Bates should about 1831 have been purchasing new tribes of cattle, and I have heard many persons assert that this arose from the loss by disease, and the decay of his old tribes; but the facts are that about 1830, when he left Ridley Hall, and settled altogether at Kirklevington, he had only a small portion of the estate in his own occupation, and adapted his herd, like any other occupier, to the size of his farm, and the numbers of his herd were very much reduced. In 1831 some of the tenants left, and Mr. Bates took the land into his own occupation, and he required about fifty head of additional stock, and to provide this he made purchases at Mr. Brown's and Mr. Parrington's sales, and he bought the Waterloo cow and others.

I, with my brothers, John and Robert Bell, had the care and management of the additional farms, and when we married we became tenants of the farms; Mr. Bates furnishing us with stock, and also purchasing and advising us in purchasing additional stock. There were then, besides the Duchesses and Red Roses, I think, eight tribes of stock at Kirklevington, and Mr. Bates partitioned them so that myself and each brother had a cow of each tribe; and we had the use of the best bulls of Mr. Bates. We very soon had very fine stock, and sold the bulls at good prices; but we were also tempted by the high prices to part with many valuable females, such as Hilpa, and, unfortunately, in several cases, we never could obtain females again, the number of bull calves being always very far in excess of the heifers; and this characteristic has continued in all the Kirklevington tribes. Whether it is from their close in-and-in breeding, or some such theory as discussed by Lord Althorp and Mr. Bates, I cannot obtain any certain information. There are many curious matters of great interest to short-horn breeders, as to colour, sexes, and the appearance and constitution of the produce, which I could enumerate from my own experience, but I fear I should tire my readers. I, however, insert the experience of another breeder of great skill and experience.

"Our attention has been drawn to an interesting point in the breeding of short-horns. Mr. Burnett, manager of the Kingscote estate and herd, informs us that two of his cows, Honey 24th and Countess of Oxford, have both produced roan calves this season to 3rd Duke of Clarence, while last season both of these cows produced white calves to the same sire. Further, it is noteworthy that of the stock begotten by this sire in 1868 none were white, in 1869 three were white, in 1870 seven were white, and in the present season none have yet been dropped white. Again, the calves by the 3rd Duke of Clarence in 1868 and 1869 were not so well formed as those of 1870, and the calves of the present season are of still more hopeful appearance. This agrees with a case recorded by Mr. Bates of one of his best stock-getters, that his first season's calves were nothing very grand, that his second season's calves were better, and that his third and fourth seasons' stock were really grand animals. Mr. Burnett considers that a bull is most valuable as a sire from three to six years old. Before that age the produce is as large, but lacks 'smartness' and 'grandeur.' That such should be the case, is in every way likely, as might be shown from analogous instances among other animals. The question of the change of colour above recorded, however, appears to belong to a different class of facts, not as easily explained. 3rd Duke of Clarence is a grandson of 7th Duke of York, a white bull, and the getter of much white stock. This alone may account for his getting white calves. Again, in the case of Honey 24th, she is by a white bull, 2nd Duke of Wetherby (21,618), and therefore not unlikely to drop a white calf to any bull, much more to 3rd Duke of Clarence. Why the 3rd Duke should get white calves one particular year and not another is interesting, and not easily explained. It is, however, a fact worth recording, and deserves the attention of breeders. We should like to know whether bulls ever do beget a large proportion of calves of a certain colour during one period of their lives, and of a different colour at another period. Very singular cases are on record of animals changing their own colour, as well as other characters, with advancing age; and it does not seem unreasonable that the offspring begotten at different periods of life may likewise vary with corresponding changes in the parent."—*Gardeners' Chronicle, May 20th,* 1871.

I may here mention that I and my brothers frequently had stock which, if trained, would have acquired the highest honours in the show yards, but Mr. Bates was so resolved not to countenance the exhibition of breeding cattle in a state of fat, that we were restrained from exhibiting. At the Scarborough show, in 1847, Mr. Bates sent his stock as a compliment, at the urgent request of Mr. Milburn, the secretary of the Yorkshire society, who was anxious to see the best stock at his exhibition, and I myself sent one untrained animal and obtained a first prize. Mr. Bates retained only about 300 acres of land in his own occupation, about 100 acres being old grass. Although the Kirklevington stock were not trained for the shows, and had been kept in their natural state on grass, the young stock generally on new grass, on the poorest parts of the estate, yet they were in such good condition, that probably few would believe that they had not been kept in the house and fed and trained in the usual manner. Mr. Bates imagined that it would be for the benefit of short-horn breeders that he should exhibit his stock in families and generations, untrained, at York, and he urged Lord Portman that there should be an exhibition for the sole benefit of breeders, before the general public were admitted, and his suggestion as to a large admission fee for the purpose was adopted by the Royal Society. When Mr. Bates showed me his letters to the council on the importance of pedigrees in breeding cattle and respecting the breeding cattle being exhibited in a proper state, I remonstrated with him as far as I could, that his suggestions would be unheeded; he however relied on the fact that he had not received any letter from the secretary that his suggestions would not be acted on, and Mr. Bates sent six cattle, and each of us his tenants one, making nine for exhibition, and four for extra stock. 2nd Duke of Oxford which had been placed before Captain Shaftoe, at Scarborough, only obtained a second prize in the local list of bulls bred in Yorkshire, and the others were unnoticed as before appears, but several cattle of Lord Feversham's breed and others also from Kirklevington bulls were commended. I had therefore no opportunity, however anxious and confident I might be, to exhibit my stock for prizes, or bring them before the public. At Kirklevington, of course the attention of visitors was too much occupied generally with Mr. Bates and his herd to give any notice of any other herd there.

On leaving Kirklevington after the death of Mr. Bates, I took several specimens of the best tribes with me into

Y

Lancashire, and I have always used bulls of the Kirklevington blood. A few years ago I removed to my present farm, as being well adapted, from the quantity of grass land, to maintain a herd in connection with a regular dairy, and with my old, and a careful selection of fresh cows, and the use of the old blood, I hoped to have had a herd not inferior to that of Kirklevington; and, in conjunction with my friends and neighbours, Messrs. Tunnicliffe and Allen, I was able to use best bulls, and had 2nd Duke of Wetherby some time in use, and in 1869 we purchased from Colonel Gunter the 8th Duke of York, then a calf, for 500 guineas, whose portrait I now insert.

We have been offered for him much more than the price of Comet.

EIGHTH DUKE OF YORK.

White, calved 15th February, 1869.

Bred by Colonel Gunter, the property of Messrs. Tunnicliffe, Allen, and Bell, by Fourth Duke of Thorndale (17750), dam Duchess 86th, by Grand Duke of Wetherby (17997).

2nd dam Duchess 70th, by Duke of Glo'ster (11382).
3rd dam do. 66th, by Fourth Duke of York (10167).
4th dam do. 55th.
5th dam do. 38th.
6th dam do. 33rd.
7th dam do. 19th.
8th dam do. 12th.
9th dam do. 4th.
10th dam do. 1st.
11th dam do. by Favourite (252).
12th dam do. by Daisy bull (186).
13th dam do. by Favourite (252).
14th dam by Hubback (319.)
15th dam by Mr. James Brown's Red Bull (97).

The Eighth Duke is growing into a good and fine bull, and his calves are good, and all those I have by him are roan. He has got very few white. Some are red, some red and white belonging to Mr. Tunnicliffe and Mr. Allen. He had been used before leaving Wetherby. At 12 months 3 weeks old his live weight was 10cwt. 7lb. He was weighed at the railway station on his arrival from Wetherby. The first calf by him was a white heifer from Duchess 103. The second calf, also a heifer, was roan from Duchess 101. These two 101 and 103 Duchesses were bought by Mr. Cochrane of Canada, and were in calf to the eighth Duke when they left Wetherby, and Mr. Cochrane describes the calves as wonderfully fine calves. In *Bell's Weekly Messenger*, May 8th, we read thus:—

Many of our readers are probably aware that Lord Dunmore is said to have bought two of the Duchess tribe in America. The following extract of a letter written to us by Lord Dunmore, dated April 28th, will explain as much of the transaction as his lordship wishes to disclose:—" I have bought the two young Duchess heifers from Mr. Cochrane, out of Duchesses 101 and 103, and I have arranged with Colonel Gunter to go on with the numbers of the Duchesses along with him; therefore I believe these two are 105 and 106. I have very good accounts of them from America, but I think I shall let them winter there, as it will harden their constitutions. The two Canadian Duchesses, are by 8th Duke of York, the white bull sold by Colonel Gunter to Messrs. Tunnicliffe, Allen, and Bell."

Duke of Hillhurst, out of Mr. Cochrane's 1000 guineas Duchess 97th, has just arrived, with 650 guineas on his head. He is a red calf, by one of the many Dukes of Thorndale.

A son of the American 11th Duchess of Geneva was bought by Mr. A. P. Davies, at Gadsby, when about a month old, for 850 guineas.

Lord Dunmore's purchase of the two heifer calves from Mr. Cochrane has just been made public. Their dams are two of

Colonel Gunter's Duchesses by the American bull 4th Duke of Thorndale. They were sold last year for (report says) 2,500 guineas, and the same price is to be paid for their American-born calves.

In speaking of the herd business during the recent Wolverhampton meeting, I have just time to notice that—

"The largest piece of business, however, done on the ground was Earl of Dunmore's purchase of five heifers of the Oxford and Red Rose tribes, from Mr. Cochrane of Canada, which are to cross with the two Duchess calves in October, and for which immense prices were paid. Duchess 97th has produced another bull calf, which has already been sold for 1000 guineas."

My son Robert Bell, who inherits the taste for improved short-horns, followed them to America, and for some time had charge of Mr. Sheldon's herd at Geneva, until he was compelled by his ill-health to return to England. This herd has since been sold and dispersed, after a very successful career.

In 1868, after the great drought, I unfortunately allowed my cattle to graze in a park with many large ornamental oak trees, and from eating the acorns I lost 30 valuable improved short-horns, and about 20 other cattle. And the foot and mouth disease, which in 1870 visited our neighbourhood, has curtailed my crop of calves, still, I hope, I have been able to preserve much of the old blood in great purity, and I can fortify the experience of Mr. Bates respecting the milking properties of the best improved short-horns. I have a tribe in which one cow, for many weeks after calving, gave regularly 9 gallons and 3 pints of milk per day. The hair and handling of the old tribe is not in the least deteriorated. When I bought the best bred cows I could meet with from other herds, I generally found that it took two crosses of Kirklevington blood to restore the true quality of the hair.

My best thanks are due to my kind friends for their sympathy and support in this, so novel an undertaking for me, and for their kindly supplying me with information of the movements of the various herds and animals. And I hope my short-comings in literary work will be overlooked and pardoned.

The importance of the production of the largest quantity of animal food from the produce of the land has immeasurably increased since Mr. Bates wrote in 1807. Remember Mr. Bates, who then called the attention of the Board of Agriculture to the subject, was not advocating the merits of improved short-horn cattle. He wished all sorts to be tested. I need not remind my readers that if the millions of the ox tribe in the British Isles were increased in productive value only one pound a year, what amount of national wealth it would produce. And, probably, Mr. Bates was not far wrong when he stated the quality of the cattle with which the land is stocked was of far more importance than the rent.

It is very consolatory to me in my declining years, and when I am going to sell my herd and retire from active life, that I have been able, however humbly and inadequately, to give the above account of the Kirklevington herd, that the public and those who possess the blood may have the benefit intended for them by Mr Bates, and I trust that the merits of the blood fully justify the prices which they now command in the market.

I may, however, as a pupil and follower of Mr. Bates, have to say of our labours—

"*Sic vos non vobis meleficites apes.*"

APPENDIX.

CATALOGUE OF THE KETTON SALE.

I have employed figures to denote the dams, which I think simplifies the pedigree, and also enables the reader to see at once the place and number of such dam. Thus d. or 1st d. means the dam, 2nd d. the granddam, &c.

COWS.

Lot.   Guineas.

1. Cherry, d. old Cherry, by Favourite, 11 years old, d. of Peeress (lot 3), Mayduke (22), and Ketton (30). Bought by J. D. Nesham, Durham ... 83
2. Kate, 4 years old, by Comet. J. Hunt, Morton, Durham .. 35
3. Peeress, 5 years old, d. Cherry, by Favourite; d. of Cecil (36). Major Rudd, Marton, Yorkshire ... 170
4. Countess, 9 years old, d. Lady, by Cupid; d. of Selina (5), Cora (12), Young Favourite (31), Young Countess (40). Major Rudd ... 400
5. Selina, d. Countess, by Favourite, 5 years old. Sir H. C. Ibbotson, Yorkshire ... 200
6. Johanna, d. Johanna, by Favourite, 4 years old. H. Witham, Yorkshire ... 130
7. Lady, d. Old Phœnix, by a grandson of Lord Bolingbroke, 14 years old; d. of Countess (4), Laura (8), Major (21), and George (32). C. Wright, Cleasby, Yorkshire ... 206
8. Laura, d. Lady, by Favourite, 4 years old; d. of Young Laura (39), and Lucilla (44). Grant, Wyham, Lincolnshire 210
9. Catherine, d. a daughter of the d. of Phœnix, by Washington, 8 years old; d. of Charlotte (42). G. Coates, for G. Parker, Sutton House, Malton, Yorkshire 150
10. Lilly, d. Daisy, by Comet, 3 years old; d. of White Rose, (46). Major Rudd. ... 410

Lot.		Guineas.
11.	Daisy, d. Old Daisy, by a grandson of Favourite, out of Venus, 6 years old; d. of Lilly (10), and Sir Dimple (33). Major Bower, Yorkshire	140
12.	Cora, d. Countess, by Favourite, 4 years old; d. of Alexander (27), and Calista (45). G. Johnston, Hackness, Yorkshire	70
13.	Beauty, d. Miss Washington, by Marske (a son of Favourite), 4 years old; d. of Albion (35). C. Wright	120
14.	Red Rose, d. of Eliza, by Comet, 4 years old; d. of Harold (29). W. C. Fenton, near Doncaster	45
15.	Flora, 3 years old, by Comet; d. of Narcissus (34). The Earl of Lonsdale	70
16.	Miss Peggy, 3 years old, by a son of Favourite. Oliver Gascoigne, Yorkshire	60
17.	Magdalene, 3 years old, by Comet, d. a heifer by Washington; d. of Ossian (28). C. Champion, Blyth, near Doncaster	170

BULLS.

18.	Comet, 6 years old, by Favourite, d. Phœnix. Wetherell, Trotter, Wright, and Charge, near Darlington	1000
19.	Yarborough, 9 years old, by Cupid, d. a daughter of Favourite. Gregson, Northumberland	55
20.	Cupid, 11 years old. Lame, was not offered for sale.	
21.	Major, 3 years old, by Comet, d. Lady. Grant	200
22.	Mayduke, 3 years old, by Comet, d. Cherry. Smithson	145
23.	Petrarch, 2 years old, by Comet, d. Venus. Major Rudd	365
24.	Northumberland, 2 years old, by Comet, d. a daughter of Favourite. Buston, Durham	80
25.	Alfred, 1 year old, by Comet, d. Venus. T. Robinson, Acklam	110
26.	Duke, 1 year old, by Comet, d. Duchess. Anthony Compton, Carham Hall, Northumberland	105
27.	Alexander, 1 year old, by Comet, d. Cora. W. C. Fenton	63
28.	Ossian, 1 year old, by Winsdor, d. Magdalene. The Earl of Lonsdale	76

Lot.		Guineas.
29. Harold, 1 year old, by Comet, d. Red Rose. Sir Lambton Loraine, Bart., Kirk Harle, Northumberland		50

BULL CALVES, NOT A YEAR OLD.

30. Ketton, by Comet, d. Cherry. Major Bower		50
31. Young Favourite, by Comet, d. Countess. P. Skipworth, Aylesby, Lincolnshire		140
32. George, by Comet, d. Lady. Walker, Rotherham		130
33. Sir Dimple, by Comet, d. Daisy. T. Lax, Ravensworth, Yorkshire		90
34. Narcissus, by Comet, d. Flora. C. Wright		15
35. Albion, by Comet, d. Beauty. T. Booth, Killerby		60
36. Cecil, by Comet, d. Peeress. H. Strickland, Boynton, Yorkshire		170

HEIFERS.

37. Phœbe, 3 years old, by Comet, d. by Favourite. Sir H. C. Ibbotson, Bart.		105
38. Young Duchess, 2 years old, by Comet, d. by Favourite. T. Bates, Halton Castle, Northumberland		183
39. Young Laura, 2 years old, by Comet, d. Laura. The Earl of Lonsdale		101
40. Young Countess, 2 years old, by Comet, d. Countess. Sir H. C. Ibbotson, Bart.		206
41. Lucy, 2 years old, by Comet, d. by Washington. C. Wright		132
42. Charlotte, 1 year old, by Comet, d. Cathelene. T. Sale, for R. Colling, Barmpton		136
43. Johanna, 1 year old, by Comet, d. Johanna. Geo. Johnston		35

HEIFER CALVES.

44. Lucilla, by Comet, d. Laura. Grant		106
45. Culista, by Comet, d. Cora. Sir H. V. Tempest, Wynyard, Durham		50
46. White Rose, by Yarborough, d. Lilly. H. Strickland, Boynton		75

Lot.   Guineas.

47. Ruby, by Yarborough, d. Red Rose. Major Bower ... 50
48. Cowslip, by Comet. The Earl of Lonsdale... 25

SUMMARY.

	Average. £ s. d.	£ s. d.
29 Cows and Heifers	140 4 7 ...	4066 13 0
18 Bulls and Calves	169 8 0 ...	3049 4 0
47 Averaged £151 8s. 0¼d.	Total...	7115 17 0

Sir H. Vane Tempest and John Hutton, of Marske, were each too late for the sale of Comet, and each is said to have been prepared to give 1500 guineas for him.

CATALOGUE OF THE BARMPTON SALE.

COWS.

1. Red Rose, 17 years old, by Favourite, d. by Ben, 2nd d. by Foljamb, 3rd d. by Hubback. Having a complaint upon her, was not offered for sale.
2. Moss Rose, 11 years old, by Favourite, d. (lot 1). Being not likely to breed again, was not offered for sale.
3. Juno, 11 years old, by Favourite, d. Wildair by Favourite, 2nd d. by Ben, 3rd d. Hubback, 4th d. sire of Hubback, 5th d. by Sir James Pennyman's bull, descended from the stock of the late Sir W. St. Quinten, of Scrampston. The Hon. J. Simpson, Badworth, Nottinghamshire, and Mr. Smith, Dishley, Leicestershire 78
4. Diana, own sister to (3). Lord Althorp 73
5. Sally, 11 years old, by Favourite, d. by Favourite, 2nd d. by Favourite. Messrs. Simpson and Smith 34
6. Charlotte, 9 years old, by Comet, d. (Catherine), bought at the Ketton sale. Mr. Brown, near Grantham, Lincolnshire 50
7. Wildair, 6 years old, by George, d. (Wildair), sister to (3). Charles Duncombe, Esq., Duncombe Park 176

Lot.	Guineas.
8. Lilly, 6 years old, by North Star, d. by Favourite, 2nd d. by Favourite, 3rd d. by Favourite. Mr. Skipworth, Aylesby, Lincolnshire	66
9. Golden Pippin, 6 years old, by North Star, d. by Favourite, 2nd d. by Favourite, 3rd d. by Favourite, from the cow that obtained the first premium at Darlington. Mr. Cattle, near York	141
10. Blackwell, 6 years old, by Willington, descended from the stock of the late Mr. Hill. Thomas Hopper, Sherburn, near Durham	31
11. Tulip, 6 years old, by George, d. by Favourite, 2nd d. by Favourite, 3rd d. by Favourite. C. Tibbets, Esq., Northamptonshire	70
12. Trinket, 6 years old, by Barmpton, d. by Favourite, 2nd d. by Favourite. Messrs. Simpson and Smith	143
13. Mary Anne, 6 years old, by George, d. by Favourite, 2nd d. by Punch. Messrs. Simpson and Smith	62
14. Louisa, 5 years old, by Wellington, d. by Favourite, 2nd d. by Favourite. Messrs. Simpson and Smith	37
15. Empress, 5 years old, by Barmpton, d. (Lady Grace) by Favourite. Charles Champion, Esq., Blyth, Northumberland	210
16. Caroline, 5 years old, by Minor, d. (Wildair) by Favourite. Henry Witham, Esq., Yorkshire	160
17. Clarissa, 4 years old, by Wellington, d. by Favourite, 2nd d. by Favourite, 3rd d. by Favourite. Mr. Robson, Holtby, Yorkshire	151
18. Young Moss Rose, 5 years old, by Wellington, d. (2). Charles Duncombe, Esq.	190
19. Venus, 5 years old, by Wellington, d. by George, 2nd d. by Favourite, 3rd d. by Punch, from a sister to the dam of the white heifer that travelled. Messrs. Simpson and Smith	195
20. Rosette, 4 years old, by Wellington, d. (1). Lord Althorp	300
21. Young Charlotte, 3 years old, by Wellington, d. (6). Mr. Thomas, Eryholm, Durham	72

Lot.	Guineas.
22. Vesper, 3 years old, by Wellington, d. by Favourite, 2nd d. by Favourite, by the dam sister to Trinket. Mr. White, Coates, near Loughborough	111
23. Nonpareil, 5 years old, by Wellington, d. (3). Lord Althorp.	370
24. Daisy, 3 years old, by Marske (the only one got by Marske), d. by Favourite, out of (5). Messrs. Simpson and Smith...	32
25. Kate, 3 years old, by Wellington, d. by Phenomenon, 2nd d. by Favourite. Henry Witham, Esq.	50
26. Amelia, 2 years old, by Lancaster, d. by North Star, 2nd d. by Favourite, 3rd d. by Punch. Mr. Maynard, Harlsey, Yorkshire	76
27. Aurora, twin sister to (26). Mr. Smith, Nottinghamshire	78
28. Princess, 2 years old, by Lancaster, d. (9). Mr. Skipworth ...	156
29. Clara, 2 years old, by Lancaster, d. (19). Mr. Thomas	190
30. Fanny, 2 years old, by Wellington, d. (5). C. Tibbets, Esq....	160
31. White Rose, 2 years old, by Wellington, d. by Wellington, 2 d. by Favourite. Mr. Smith ...	51
32. Ruby, 2 years old, by Wellington, d. (1). Mr. Robson	331
33. Lavinia, 2 years old, by Lancaster, d. (18). Mr. Robson ...	105
34. Hebe, 2 years old, by Jupiter, d. (8) Mr. Thompson, Scremerston, near Berwick	90
35. Jesse, 2 years old, by Wellington, d. from the stock of the late Mr. Hill. Mr. Hutchinson...	43
36. Jewel, 2 years old, twin sister to (35). Mr. Brown...	59

HEIFERS, FROM SIX TO TWELVE MONTHS OLD.

37. Violet, by North Star, d. by Midas, 2nd d. by Punch. Mr. Skipworth	48
38. Sweetbrior, by North Star, d. (23) Mr. Maynard ...	145
39. Snowdrop, by Wellington, d. (11). Mr. Thompson, Stockton, Durham	71
40. Cowslip, by Wellington, d. by Favourite, 2nd d. by Punch. Mr. Layton, North Willingham, Lincolnshire	54
41. Lady Anne, by Wellington, d. by George. 2nd d. (3) Mr. Barnes, Eggleston	100

Lot.	Guineas.
42. Flora, by Lancaster, d. (5) Mr. Thompson	47
43. Cleopatra, by Lancaster, d. by George, 2nd d. by Favourite, 3rd d. by Punch. Mr. Barnes	133
44. A heifer, by Lancaster, d. (17) Calved 26th September. Mr. Robson	52
45. By Lancaster, d. (12) Calved 28th October. Mr. Wiley, Yorkshire	56
46. By Wellington, d. by Wellington. Calved 20th October. Messrs. Simpson and Smith	28
47. By Lancaster, d. (13) Calved 16th November. Mr. Cattle	42
48. By Lancaster, d. (14) Calved 20th November. Mr. Smith.	38
49. By Barmpton, d. (15) Charles Champion, Esq.	100
50. By Barmpton, d. (20) Mr. Robson	123
51. By Barmpton, d. (6) Major Rudd	55
52. By Barmpton, d. (12) Messrs. Simpson and Smith	110
53. By Barmpton, d. by Cleveland, 2nd d. by Comet, 3rd d. by Favourite. Mr. White	80

BULLS.

54. Marske, 12 years old, by Favourite, d. by Favourite, 2nd d. by Favourite. Mr. Maynard	50
55. North Star, 11 years old, by Favourite, d. by Punch. T. Lax, Esq.	72
56. Midas, 10 years old, by Phenomenon, d. (1) Mr. Wiley	270
57. Barmpton, 8 years old, by George, d. (2) Being lame, was not sold.	
58. Major, 5 years old, by Wellington, d. by Phenomenon, 2nd d. by Favourite, 3rd d. by Favourite. Mr. Brooks, Laceby, near Grimsby, Lincolnshire	185
59. Lancaster, 4 years old, by Wellington, d. (2) Messrs. Simpson and Smith	621
60. Baronet, 3 years old, by Wellington, d. (1) Being engaged, was not put up.	
61. Regent, 3 years old, by Wellington, d. by Windsor, 2nd d. (1). Lord Althorp	145

BULL CALVES.

Lot.		Guineas.
62.	Diamond, 1 year old, by Lancaster, d. (19) Mr. Donaldson, Harburn House, near Durham	102
63.	Albion, rising 1 year old, by Lancaster, d. by Wellington, 2nd d. by Favourite, 3rd d. by a son of Favourite. Mr. Russell, Brancepeth Castle	140
64.	Harold, rising 1 year old, by Wellington, d. (7) Mr. Whittaker, near Otley, Yorkshire	201
65.	Pilot, rising 1 year old, by either Major or Wellington, (being bulled by both), d. (1) Mr. Booth, Killerby, near Darlington, Durham	270
Amount of First Day's Sale...		£7852 19 0

Coates's Herd Book had originally been intended to embrace a sort of history of the animals inserted in it. I now notice the most remarkable animals, without the formal pedigrees, except in a few cases, with the remarks of Coates mentioned above. G. Coates had, most probably, seen all the animals, which he thus notices in 1822. B denotes bull, and C. C., Charles Colling, and R. C., Robert Colling, the numbers are those in the Herd Book.

(2) Achmet. Sold to Mr. Morley, 70 guineas. Hollon, whose stock were by Dalton Duke and Mr. Hills, of Blackwell. Hollon (313), either by Dalton Duke or a B. out of Chapman's cow.

(14) Albion, by Comet, d. (Beauty) by Marske, 2nd d. Miss Washington. Mellow, not good, but his stock generally good.

(19) Ralph Alcock's B. Remarkable for good handling and lively looks.

(32) Richard Barker's B., (see page 54), sire of Foljamb. Good size and symmetry, hard handle.

A gentleman informs me that Wetherall, Wynyard, and Wiley, all shortly before they died gave him the same account of this bull as given by Mr. Bates.

(76) Blaize. Very good, had a dark nose.

(86) Bolingbroke (or Lord B.) Blood red, with a little white. Calved November 12th, 1788, bred by C. C., by Foljamb, d. (Young Strawberry, bred by Mr. Maynard) by Dalton Duke, 2nd d. (Favourite) bred by Mr. Maynard, by Alcock's B., 3rd d. by Jacob Smith's B., 4th d. by Jolly's B. "The best bull G. C. ever saw."

(97) James Brown's old red B. Good fore-quarters, and handle, huggins and rumps not good, strong thighs, excellent getter.

(120) Cecil. Sides too flat.

(107) Candour. Bad.

(123) Charges' roan or grey B., bred by Mr. Charge, by Favourite. Fair.

Colling's red B., 1819, bred by Simpson, by Lancaster, d. Juno, &c., &c. Very bad. 'Tis a pity no one will tell Mr. Simpson.

(155) Comet, red and white roan, calved 1804, by Favourite, d. (Young Phœnix) by Favourite, 2nd d. (Phœnix) by Foljamb, 3rd d. (Favourite, bred by Mr. Maynard) by Alcock's B. I never saw his equal.

(188) Dalton Duke, brindled red and white, bred by J. Charge, by Mr. Dobson's B., d. by Studley B. Wide back, great substance. He was sold to Maynard and Duke Wetherall, at 2 years old, for 50 guineas.

(209) Diamond. Duncombe. Good, if knuckles not too large.

(218) Dobson's B., bred by Mr. Dobson, by a B. bred by Jacob Smith. Sold to Mr. Charge. A good B., and his stock superior.

(226) Duke. Very good, except broad top of his shoulders.

(237) Eclipse. His stock bad.

(239) Eadlesthorpe. Very moderate, knuckles bad.

Favourite (252), light roan, calved in 1793 or 94, bred by Mr. C. C., by Bollingbroke (86), 1st d. (Phœnix) by Foljamb (263), 2nd d. (Favourite) by R. Alcock's B. (19), 3rd d. by Jacob Smith's B. (608), 4th d. by Jolly's B. (337). Favourite was a large beast, fine bold eye, low back, body down, other parts very good. Was used by Mr. Charge, 1798 and 1799.

(263) Foljamb, white, with a few red spots, calved 1787, bred by Mr. C. C., by Richard Barker's B., d. (Haughton) by Hubback. Foljamb, useful thick beast, handle good, dark face,

Sold by G. C. to Mr. Foljamb, at about twenty months old, for 50 guineas. I heard Mr. C. C. say this beast did him most good.

(273) George, calved 1802, (Mason) by Favourite, d. (bred by Col. Simpson), by Ben, 2nd d. bought of C. C., which he bought of John Newby.

Grey B. (not in Herd Book), bred by R. C, by white B., d. own sister to the dam of R. C.'s Red Rose. Grey B. handle like Hubback.

(290) Harold, 1809, bred by C. C., by Windsor, d. (Red Rose) by Comet, 2nd d. Eliza; Eliza's d. a heifer bought by C. C. of Mr. Bates. Sold to J. L. Loraine, 50 guineas.

(291) Harold, 1818. Harold good, but handle hard.

(319) Hubback. Pedigree as set out by Mr. Bates. The following account of the dam of Hubback was given to the author by the undermentioned person. "I remember the cow which my father bred. That was the dam of Hubback. There was no idea then that she had any Mixed or Kyloe blood in her. Much has been said lately that she was descended from a Kyloe; but I have no reason to believe, nor do I believe that she had any mixture of Kyloe blood in her.—JOHN HUNTER, Hurworth, near Darlington, July 6th, 1822." Hubback, head good, horns small and fine, neck fine, breast well-formed and firm to the touch, shoulders rather upright, head or girth good, loins, belly and sides fair, rump and hips extraordinary, flank and twist wonderful. Mr. Maynard sent cow to Hubback at Ketton, 1787, B. C. C. used him 2 or 3 years, and sold him and Foljamb in the same year.

(314) Jupiter, by Comet, d. (Beauty, bred by C. C.) by Marske, 2nd d. Miss Washington. Fair, but neck very large. (Brother to Albion).

Lakeland's Bull. Great size, good back. Sold into Northumberland.

(360) Lancaster. Brother to Young Moss Rose. Good. Rumps indifferent.

(365) Lawnsleeves, by C. C.'s Surplice, d. by George, 2nd d. by Simon, out of Smith's old cow. Bred at Halnaby.

(397) Major, by Comet, d. Lady, by Grandson of Bolingbroke. Fair.

(413) Mars. Hard handle, short hair. Altogether bad.

(415) Marshall Beresford. Own brother to Lily and Sir Dimple. Blacklocks space large, not good.

Maximillian, 1823. J. W. Got bad stock.

(418) Marske, died 1822. Hips moderate, good flesh.

(435) Midas. Shoulders too upright and coarse, mellow and good handle.

(451) Nero. Good.

Newton. Charge. Very good, except the tail too high.

(471) Orlando. Hutchinson. Bad, coarse hind quarters, hair bad, cloven back.

Northumberland. Harrison. A fair B., knuckles a little too large.

Oliver, bad constitution. Entered in pencil only.

(486) Patriot. G. Coates. Sold for 500 guineas.

Percy, 1821. J. Booth. Roan, by Pilot, d. (Young Rosette) by Albion, &c., &c. Shoulders and crop indifferent.

(488) Petrach. Horn too short, good B. Castrated in 1818.

(491) Phenomenon, bred by R. C., by Favourite, d. by Ben, 2nd d. by Hubback, 3rd d. by Snowdon's B., 5th d. by Sir J. Pennyman's B. Large and good. Brother to Wildair.

(496) Pilot. Lot 65 at Barmpton. Bad crops.

(514) Regent. Horns up, good.

(580) Remus. Wright. Brother to Lauristine. Sold to Mr. Hutton for 110 guineas. Handle not good.

(588) Robson's bull, by Masterman's bull. The dam gave 36 quarts (ale measure) a day on grass only for months.

(603) Sir Leoline. Dark, red flesh. Hutchinson. Liry handle. Must be from the black tribe.

(605) Sir Oliver. Good.

(612) Snowdon's bull, sire of Hubback. Cow came from Sir W. St. Quinten, which came to Sir James P. in 1795. A neat good beast. Mr. Osler's father offered 50 guineas for him.

(627) Studley white B. Mr. Charge used this B. for five or six years. A good B., with extra good and large fore-quarters.

Studley, roan, got a Premium at Bedale, 1820. Good.

(672) Warlaby, roan. Good, except shoulders. Got a premium at Doncaster. The sire of Judson's B.

(673) Warrior. Handle a little hard.

(674) Washington, by Favourite, d. (Lady) by grandson of Bolingbroke, brother to Laura.

(680) Wellington. R. C. Very good, ribs excellent.

(689) Western Comet, descended from the Studley B. Rich handle, and all his stock good.

(151) White B., R. C., by Favourite, d. by Hubback. In 1840, R. C. said that he had nothing but this White B., that went direct from Hubback to Favourite, and C. C. admitted that he had nothing so descended, except the Daisy B.

(698) Windsor. Rich handle, much like Western Comet, but head larger. With Mr. Hustler, 1808-1809.

(703) Wynyard. Sold to Mr. Milner, 210 guineas.

(469) O'Callaghan's son of Bolingbroke, red and white, brindled, bred by Colonel O'Callaghan, of Heighington, near Darlington, by Bolingbroke, d. (a red polled Galloway Scotch cow). This cow and another of the same breed were purchased of D. Smurthwaite, near Northallerton, by George Coates, who sold them to Colonel O'Callaghan. O'Callaghan's son of Bolingbroke, when a few days old, became the property of C. Colling, and was the sire of grandson of Bolingbroke.

Beauty, roan, calved 1806, bred by C. C., by Marske, d. Miss Washington.

Produce and Name.	Sire.
1810 Light roan, Albion (14), Booth	Comet.
„ ...Jupiter (343), Wright	Comet.

Miss Washington does not appear in the Herd Book, but I suppose she was by Washington (674), dam unknown.

Fortune (p. 316), calved 1793, bred by C. C., by Bolingbroke, d. by Foljamb, 2nd d. by Hubback, bred by Mr. Maynard.

Produce and Name.	Sire.
1796...B.	Favourite.
1797...Gaudy	Favourite.
1798...Irishman	Styford.

Produce and Name.	Sire.
1799...Nell	White Bull.
1800..Cripple	Irishman.
1802...Trunnell	Favourite.
1803...Tidy	Union.
1804...St. John	Favourite.
1806...S.	Chilton.

Favourite or Lady Maynard (p. 307), red roan, bred by Mr. Maynard, by R. Alcock's Bull, d. by J. Smith's Bull, 2nd d. Strawberry, by Jolly's Bull.

Produce and Name.	Sire.
——..Young Strawberry	Dalton Duke.
——...Miss Lax	Dalton Duke.
——...Phœnix	Foljamb.
——...Lady Maynard's Bull	Lame Bull.
1796...Mason's White Bull	Bolingbroke or Favourite.

Old Johanna (p. 343), bred by C. Colling, by C. C.'s Lame Bull. Produce—1793 or 4, Red and White, grandson of Bolingbroke, by O'Callaghan's son of Bolingbroke.

Lady, red and white, calved in 1796. Bred by C. Colling, by grandson of Bolingbroke, d. (Phœnix) by Foljamb, &c.

Produce and Name.	Sire.
——...Roan, Washington (674)	Favourite.
1801...Red, Countess	Cupid.
1806...Red and White, Laura	Favourite.
1807...Major (397)	Comet.
1810...George (276)	do.
——...Sir Charles (592)	do.

I leave my readers to "scan the results in so many fairs and pastures. We may well feel that short-horns have repaid all the money, thought, and labour, which have been bestowed upon them. Still, in one way only can their supremacy be made permanent,—by always keeping in mind the rule by which our first breeders have been guided, that 'a good beast must be a good beast however he has come; but, that, it is to pedigree alone that we can trust for succession.'"*

* Dixon, Royal Agricultural Society's Journal, vol. 1, page 329. Saddle and Sirloin, page 175.

FINIS.

ERRATA.

Page 38, Line 1, for crossed, read *crosses*.
,, 38 ,, 23, for my, read *any*.
,, 39 ,, 11, for by, read *of*.
,, 45 ,, 4, for was, read *saw*.
,, 54 ,, 20, for old, read *Colonel*.
,, 56 ,, 16, for tribe, read *tribes*.
,, 59 ,, 9, for these, read *they*.
,, 77 ,, 6, for take, read *taken*.
,, 83 ,, 22, for holster, read *bolster*.
,, 90 ,, 11, for forward, read *forwarder*.
,, 90 ,, 25, for Duchess, read *Duchesses*.
,, 106 ,, 22, for grand, read *god*.
,, 112 ,, 16, for Prudhoe, read *Sandhoe*.
,, 120 ,, 17, for Culley, read *Colling*.
,, 123 ,, 6, for persuming, read *presuming*.
,, 125 ,, 27, for meeting, read *meetings*.
,, 139 ,, 15, for Holkam, read *Holkham*.
,, 151 ,, 19, for in 1812, read . *In* 1812.
,, 152 ,, 21, for freey, read *freely*.
,, 172 ,, 24, for Crathome, read *Crathorne*.
,, 177 ,, 5, dele *and the*.
,, 195 ,, 8, for dam, read *son*.
,, 195 ,, 9, for hornless, read *no hornless*.
,, 197, last line, for thei, read *their*.
,, 206, line 16, for size, read *sire*.
,, 208 ,, 23, for sock, read *stock*.
,, 208 ,, 23, for Firciso, read *Furioso*.
,, 209 ,, 28, for he, read *Mr. Bates*.
,, 210 ,, 16, for cross, read *cress*.
,, 211 ,, 16, for first g., read *by*.
,, 217 ,, 4, for at, read *to*.
,, 217 ,, 16, for blank, read *black*.
,, 243 ,, 20, for poetry the, read *poetry and*.
,, 244 ,, 6, for premium, read *pre-eminence*.
,, 272 ,, 16, for Bunse, read *Bunsen*.
,, 276 ,, 22, for work, read *mark*.

NEWCASTLE-UPON-TYNE:

PRINTED AT THE "NORTH OF ENGLAND FARMER" OFFICE,
CLAYTON STREET.

DUCHESS XCIV. & CALF.

DUCHESS XXXIV.

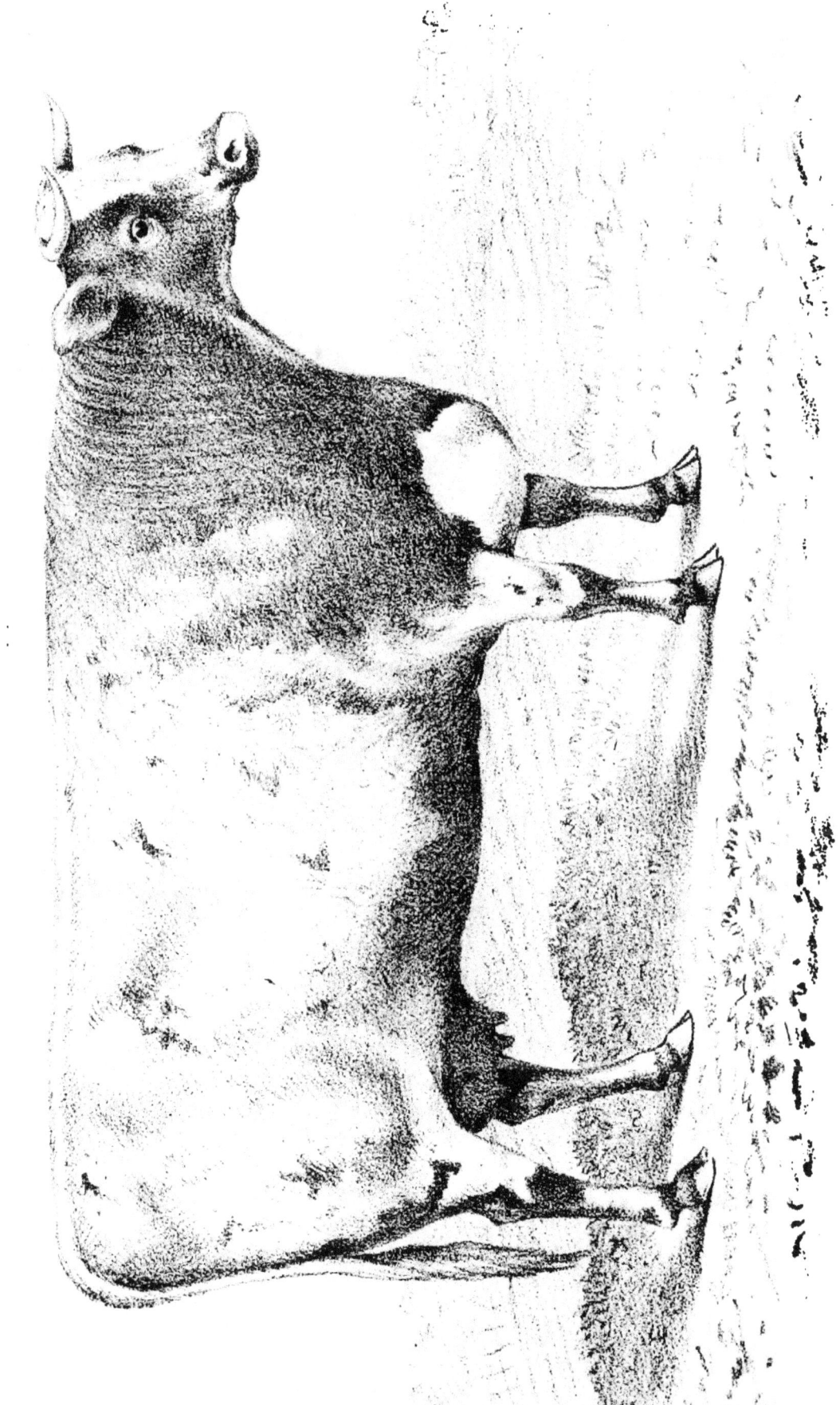

ELVIRA.

DUCHESS DAM or KETTON the 1st.

www.ingramcontent.com/pod-product-compliance
Lightning Source LLC
Chambersburg PA
CBHW082320220526
45470CB00008B/2364